Kaye Mash

How Invertebrates Live

ELSEVIER PHAIDON

Credits
to Photographers

Artists and photographers are listed alphabetically with their agents' initials, where applicable, abbreviated as follows:
(AFA) Associated Freelance Artists Ltd.;
(Ardea) Ardea Photographics;
(NSP) Natural Science Photos;
(Res) Bruce Coleman Ltd.

Aquatic (NSP) 38(bottom)
D. Bartlett (Res) 49
J. Burton (Res) 27(bottom) 30(bottom) 34 46(top) 47 56 57 61 62 80(bottom) 81 96 101 103 122
D. Corke 29
W. Curth (Ardea) 141
C. Doeg (NSP) 153
C. Doncaster 68
Elsevier Nederland B.V. 17 24(top) 25(top) 27(top) 28(bottom) 30(top) 32 35 39(top) 46(bottom) 51(bottom) 58(top) 64 70 71 72 73 74 80(top) 82(2) 84(top) 90 94 98 102(top) 109(top) 112(2) 116 117(2) 124(top) 127 130(top) 136(top) 149(bottom) 152 155
Fotosub 124(bottom) 134
J. A. Grant (NSP) 19
W. Harstrick (Res) 54
J. S. Hillock (NSP) 91 100

J. Holt (NSP) 25(bottom) 123
D. A. Kempson 36
R. Kinne (Res) 21(top) 93 148
G. Kinns 51(top)
A. Leutscher (NSP) 121 138
D. B. Lewis (NSP) 6 15 78 102(bottom) 132 149(top)
Dr. P. A. Morris 14 58(bottom) 79 87 88 97 118(bottom) 150(2) 154
R. K. Murton (Res) 119
Oxford Illustrators 8 9 10 11 12 21(bottom) 33 85 106 125
L. E. Perkins (NSP) 89 95
G. Pizzey (Res) 20 67
Dr. T. B. Poole 76
A. Power (Res) 69 129
Dr. M. A. Sleigh 24(bottom)
Dr. R. Teunissen 50
The Zoological Society of London 142(bottom)
P. H. Ward (NSP) 38(top) 52(2) 97 142(top) 143
Dr. D. P. Wilson 2 16(bottom) 22(2) 23 26 31 39(bottom) 40 41 42 43 44 45 48 53 60 63 66 75 84(bottom) 92 104(3) 105 107 108(2) 109(bottom) 110(2) 111(2) 113 114(2) 115 118(top) 126(2) 128 130(bottom) 131 133 135 137 139 140 147
A. van den Nieuwenhuizen 16(top) 18 28(top) 37 59 65 77 83 136(bottom) 145 146 151

Elsevier-Phaidon,

An imprint of Phaidon Press Ltd.
5 Cromwell Place, London SW7 2JL.

First published 1975
Planned and produced by
Elsevier International Projects Ltd, Oxford
©1975 Elsevier Publishing Projects SA, Lausanne.

ISBN 0 7290 0024 9

Filmset by Keyspools Limited, Golborne, Lancashire
Printed in Belgium

oto Montage
re Cnidaria

PP. 10 - 47

How Invertebrates Live

How Animals Live

edited by

Peter Hutchinson

*a series
of volumes
describing
the behaviour
and ecology
of the animal
kingdom*

VOLUME 4

Contents

The Invertebrates

Ninety-five per cent of all animals are invertebrates. Some are well known because they are common, like earthworms, Sea anemones and snails, while others are appreciated as food, like oysters, crabs and lobsters. The vast majority, however, are rather obscure and known only to the zoologist. They are nevertheless of considerable interest. Many have microscopic larvae, and it is a revelation to see these beautiful animals alive and often it is a great surprise too, when looking down a microscope for the first time, to find that ordinary objects such as pieces of moss or pond weed teem with a multitude of bizarre microscopic animal types.

The largest groups of invertebrates are the arthropods or joint legged animals and the molluscs, a group including snails and clams, which contain 852,000 and 128,000 species respectively. This contrasts with a mere 4,000 species for all the mammals, the group to which man belongs, and means that three-quarters of the total number of animal species known are arthropods. If success is rated in terms of numbers, the vertebrates, which include the fish, amphibians, reptiles, birds and mammals, fall some distance behind both arthropods and molluscs with only 43,000 species altogether. Insects account for most of the arthropod species. Not only are there a great many species, some are so numerous that they have more members than the whole of the human population!

Because the invertebrates constitute such a large part of the animal kingdom, it is not surprising that they include an incredibly diverse assemblage of types, ranging from those composed of a single cell, to the giant octopuses and squids. Many of these types appear to be thoroughly bizarre, for example, some seem to manage without brains or even heads; the sluggish clams, tenacious barnacle and flower-like Sea anemone are examples of invertebrates that have either lost their heads (in an evolutionary sense) or have never had them. Amongst the 'alien' features shown by other invertebrates are their peculiar body shapes; plant-like forms built on a radial or circular plan like corals and sponges, globular floating or free wheeling forms like planktonic larvae, flat animals, curled animals and tubular animals. Many invertebrates, such as flatworms, leeches and roundworms, have no limbs and wriggle, swim and burrow like snakes, or have

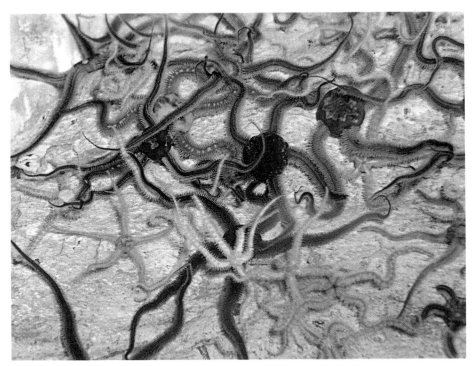

Invertebrates such as these brittlestars often appear bizarre and quite unlike the vertebrate animals with which we are more familiar. They are, nevertheless, highly evolved organisms which are as well adapted to their environment as are better known animals.

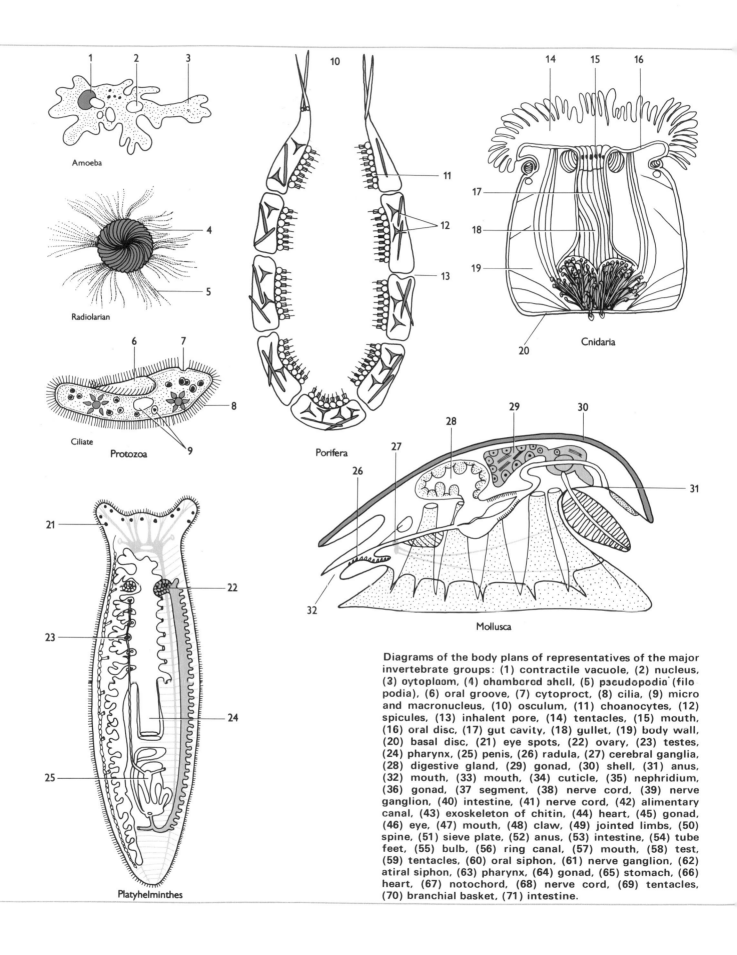

Amoeba

Radiolarian

Ciliate

Protozoa

Porifera

Cnidaria

Mollusca

Platyhelminthes

Diagrams of the body plans of representatives of the major invertebrate groups: (1) contractile vacuole, (2) nucleus, (3) cytoplasm, (4) chambered shell, (5) pseudopodia (filopodia), (6) oral groove, (7) cytoproct, (8) cilia, (9) micro and macronucleus, (10) osculum, (11) choanocytes, (12) spicules, (13) inhalent pore, (14) tentacles, (15) mouth, (16) oral disc, (17) gut cavity, (18) gullet, (19) body wall, (20) basal disc, (21) eye spots, (22) ovary, (23) testes, (24) pharynx, (25) penis, (26) radula, (27) cerebral ganglia, (28) digestive gland, (29) gonad, (30) shell, (31) anus, (32) mouth, (33) mouth, (34) cuticle, (35) nephridium, (36) gonad, (37 segment, (38) nerve cord, (39) nerve ganglion, (40) intestine, (41) nerve cord, (42) alimentary canal, (43) exoskeleton of chitin, (44) heart, (45) gonad, (46) eye, (47) mouth, (48) claw, (49) jointed limbs, (50) spine, (51) sieve plate, (52) anus, (53) intestine, (54) tube feet, (55) bulb, (56) ring canal, (57) mouth, (58) test, (59) tentacles, (60) oral siphon, (61) nerve ganglion, (62) atiral siphon, (63) pharynx, (64) gonad, (65) stomach, (66) heart, (67) notochord, (68) nerve cord, (69) tentacles, (70) branchial basket, (71) intestine.

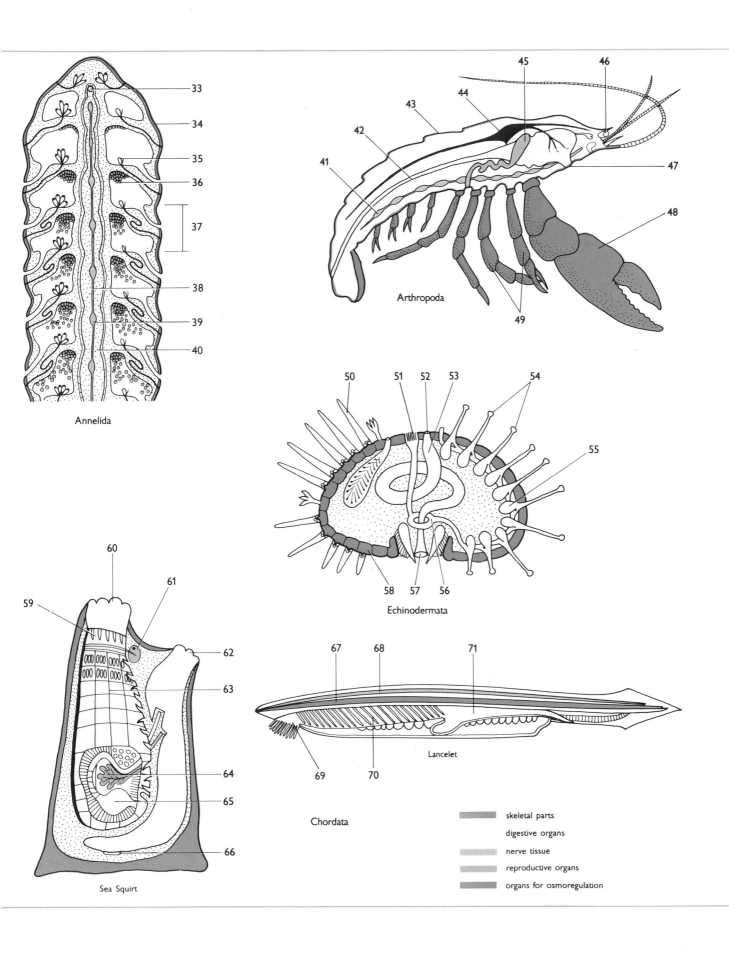

33

34

35

36

37

38

39

40

Annelida

45   46

44

43

42

41   47

                                                                        48

Arthropoda   49

50   51 52 53   54

                                                55

58 57 56

Echinodermata

60

61

59

                                    62

                                    63

                                    64

                                    65

                                    66

Sea Squirt

67   68   71

69   70

Chordata

Lancelet

skeletal parts

digestive organs

nerve tissue

reproductive organs

organs for osmoregulation

Classification of the Invertebrates

SUBKINGDOM	PHYLUM	CLASS	EXAMPLES
PROTOZOA	PROTOZOA	MASTIGOPHORA	Flagellates
		SARCODINA	Amoebae
		SPOROZOA	Parasites
		CILIATA	Ciliates
PARAZOA	PORIFERA		Sponges
MESOZOA	MESOZOA		Parasites of doubtful affinity
METAZOA	CNIDARIA (COELENTERATA)	HYDROZOA	Obelia, hydra, Sea firs
		SCYPHOZOA	Jellyfishes
		ANTHOZOA	Sea anemones, corals
	CTENOPHORA		Comb jellies
	PLATYHELMINTHES	TURBELLARIA	Flatworms, planaria
		MONOGENEA	Fish parasites
		CESTODA	Tapeworms
		DIGENEA	Flukes
	NEMERTINA		Ribbonworms
	ACANTHOCEPHALA		Thorny headed worms
	ROTIFERA		Wheel animals
	NEMATODA		Roundworms
	KINORHYNCHA		
	GASTROTRICHA		
	NEMATOMORPHA		Hairworms
	PHORONIDA		Horseshoe worms
	BRYOZOA (ECTOPROCTA)		Moss animals
	ENTOPROCTA		

SUBKINGDOM	PHYLUM	CLASS	EXAMPLES
	BRACHIOPODA		Lamp shells
	PRIAPALIDA		Proboscis worms
	MOLLUSCA	MONOPLACOPHORA	Neopilina
		POLYPLACOPHORA	Chitons
		GASTROPODA	Snails, slugs, Sea snails
		SCAPHOPODA	Tusk shells
		BIVALVIA	Clam, oysters, scallops
		CEPHALOPODA	Octopus, squid, cuttlefish
	ANNELIDA	POLYCHAETA	Ragworms, tubeworms
		OLIGOCHAETA	Earthworms
		HIRUDINEA	Leeches
	SIPHUNCULOIDA		Peanut worms
	TARDIGRADA		Water bears
	ONCHYOPHORA		
	ARTHROPODA	CRUSTACEA	Crabs, lobsters, prawns
		INSECTA	Insects
		ARACHNIDA	Spiders, ticks, mites
		DIPLOPODA	Millipedes
		CHILOPODA	Centipedes
		MEROSTOMATA	King crabs
	CHAETOGNATHA		Arrow worms
	POGONOPHORA		
	ECHINODERMATA	CRINOIDEA	Sea lilies
		ASTEROIDEA	Sea stars
		OPHIUROIDEA	Brittlestars
		ECHINOIDEA	Sea urchins, Sand dollars
		HOLOTHUROIDEA	Sea cucumbers
	HEMICHORDATA		Acorn worms
		SUB PHYLUM	
	CHORDATA	UROCHORDATA	Sea squirts, salps
		CEPHALOCHORDATA	Lancelets

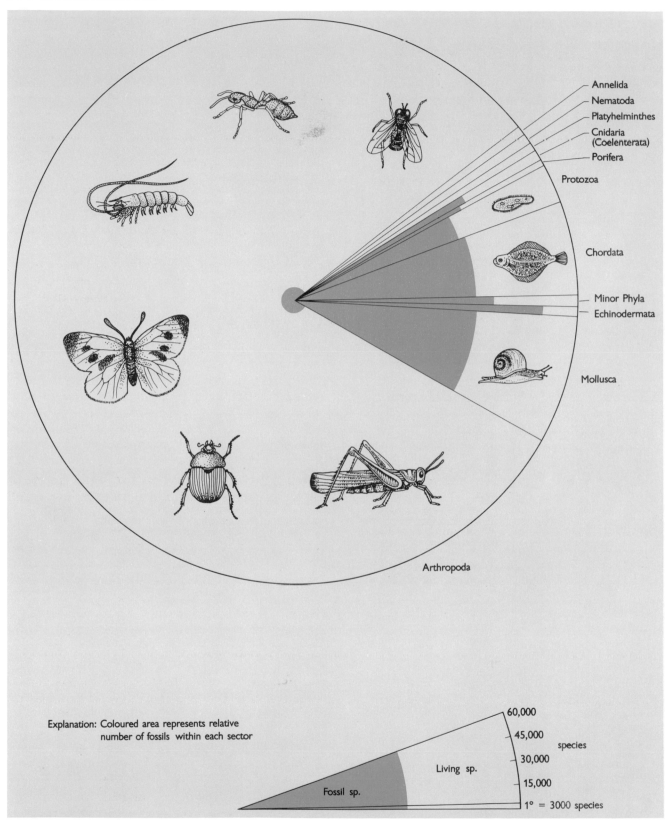

Annelida
Nematoda
Platyhelminthes
Cnidaria
(Coelenterata)
Porifera
Protozoa
Chordata
Minor Phyla
Echinodermata
Mollusca
Arthropoda

Explanation: Coloured area represents relative
number of fossils within each sector

60,000
45,000
30,000
15,000
species
Living sp.
Fossil sp.
1° = 3000 species

The importance of the invertebrates in the animal kingdom is demonstrated in this diagram which shows that the arthropods (lobsters, crabs and all the insects) account for at least three quarters of the known species. The fact that very few fossil arthropods are known is the result of their not being easily preserved.

reduced their limbs to a few bristles as in earthworms. Where limbs are present these often occur in peculiar numbers and arrays. Sea stars frequently have arms in fives or groups of five, insects have six legs, spiders have eight, marine bristleworms have hundreds of legs and so do millipedes. Some of the lower invertebrates such as Sea anemones, corals and flatworms have guts with only a mouth and no anus and have to regurgitate waste. But however strange they are, all invertebrates are perfectly adapted to their life style and are in fact the products of millions of years of evolution.

The Classification of Invertebrates. The diversity of form displayed by the invertebrates has led to their being classified into three sub-kingdoms: the Protozoa or single celled animals; the Parazoa or sponges, in which the cells of the body are loosely integrated; and the Metazoa which are the truly multicellular animals. Zoologists have had to split the invertebrates further into 29 major groups (phyla) each of which is comparable to the Chordata, the phylum which contains the vertebrates, or animals with backbones. The different groups of invertebrates are often at least as different from one another as they are from the vertebrates.

The brief survey of the invertebrates presented in this book aims to some extent to be an exposé of alternative life styles, ranging from filter feeding sponge to parasitic sponger, from burrowing worm to floating transparent jellyfish, from animals made up of many fused bodies to solitary individuals. Invertebrate lives are not easily understood and often seem quite alien to our own vertebrate existence, sometimes more like science fiction than reality. It is amazing that, for instance, invertebrates called protozoans can perform all the functions of life in a single cell; that sponges coordinate their albeit limited activities quite well without any obvious nervous system; and that many invertebrates have larvae quite different from the adult stage leading very different lives, which have to metamorphose to transform into adults.

There are of course books on invertebrate animals but these are usually written in technical language mainly for zoologists, which does tend to obscure the fascinating lives that these animals lead. This book attempts to introduce the basic features and ways of life of members of all the major groups of invertebrates except the insects which, being such an enormous group, form the subject of another book in this series. One point, however, must be made at the outset. Many of the animals discussed here do not have common names, and scientific names have therefore been employed. Similarly, a small number of technical words are unavoidable but, when used, they have been carefully explained.

This book starts with a classification of living invertebrates and then introduces the body plans of the major groups. The following chapter on skeletons and movement shows how the basic organization of the body is intimately related to locomotion as diverse as amoeboid movement and flying. Later, we investigate the ways in which invertebrates coordinate their often complex behavioural activities. As invertebrates apprehend the world through often very different kinds of sense organs with different sensitivities from our own, there is a section on sense organs and perception. The way invertebrates feed and the food chains they themselves participate in is dealt with and there is a special chapter on animals that live in association, in particular parasites and their adaptations to their often seemingly impossible modes of life. Other topics covered are courtship, reproduction, parental care and larval lives and special consideration is given to the ways in which many invertebrates survive adverse conditions by remaining in a state of suspended animation. The advantages of colonial life and the ways in which colonies arise are also investigated. So too is that strange phenomenon known as polymorphism, in which an animal may be a chimaera of fused individuals of completely different appearance. Colour plays as much importance in the lives of invertebrates as in vertebrate animals in courtship and other kinds of behaviour, as a warning mechanism for poisonous forms and in camouflage. The role of colour in invertebrate lives and the biology of coloured respiratory pigments forms the subject of another chapter and is considered together with venomous invertebrate animals. Invertebrate animals of course exert a profound influence on man and are the basic fabric of all life. They feature in all the complex food webs of higher animals and are sometimes cultivated or captured to be eaten directly by man, like oysters and lobsters. Plant pests, pests of stored products and parasites exert a powerful destructive influence on agriculture, the economy and the health of man and his domestic animals. Other invertebrates, such as earthworms and pollinating insects, are beneficial. There are signs that marine invertebrates such as sponges and corals may increasingly yield useful drugs. These and other aspects of man's relationship to invertebrates are considered in the final chapter.

Skeletons and Movement

Just as man is remembered after death by his arte-facts and architecture, so many invertebrate animals leave as monuments to their passing their hard shells and skeletons. The most familiar of these skeletons are the sea scoured shells strung out on the beach along the high water line. Many shells have long been collectors' items, others like the Money cowrie which was used for currency in West Africa as recently as 100 years ago and the ear-shaped abalone shell used for stamping out 'mother of pearl' buttons, have found more practical uses. The less conspicuous shells of minute marine protozoans such as the foraminiferans make their mark in numbers and a constant rain of their calcareous skeletons falling on to the sea bed has led to the deposition of chalk and limestone. The hardened siliceous skeletons of radiolarian proto-zoans have produced flint and chert in a similar

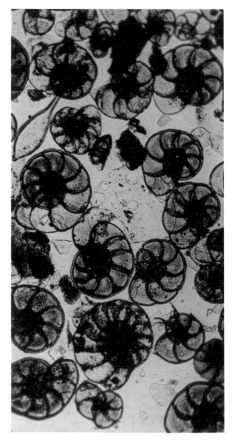

Shells of dead foraminiferan protozoans are responsible for large chalk deposits such as the cliffs of Dover. The shells have a number of chambers arranged in a spiral.

way. It is sobering to imagine how many millions and millions of protozoan skeletons led to the formation of a single stick of chalk.

Functions of Skeletons. Just as there are many kinds of animal skeleton, so skeletons have many functions. Hard skeletons may be worn inside (endoskeleton) or outside (exoskeleton) the body. Even when there is an exoskeleton, however, this may send processes into the tissues for muscle attachment as in the cuticle of lobsters and insects. Not all animals have hardened skeletons; some simply pump themselves up with fluid, giving them a kind of rigidity that muscles can act on. The roundworms and earthworms have fluid skeletons surrounded by muscles and forming an incom-pressible core to the body. Skeletons give the body shape and support and this is particularly im-portant in land animals where there is no water to buoy up the body. They also either provide a firm attachment point for muscles or help the muscles to stretch back after contraction by their built in elasticity or incompressibility. Hard skeletons like shells and cuticles are protective against physical and mechanical damage to the body and against predators. Animals living in the surf zone of a rocky shore give a good example of this and forms like the limpet, Dog whelk and mussel have very thick resistant shells. This is partly to allow them to withstand the battering of waves and stones, but also to offer protection against drying out when the tide retreats and to confer some immunity against the attentions of predators.

Some kinds of skeleton aid attachment, like the hooks and spines of parasites, the byssus threads of bivalves which form the so called 'beard' of mussels and the tufts of silica spines anchoring the Glass rope sponge to the sandy seabed. More unusual functions of skeletons are as flotation devices like cuttlebone or as cutting plates in wood and rock-boring bivalves. The 'shipworm' *Teredo* has long siphons and small sharp blade-like shell valves used to carve circular tunnels in submerged wood as the body is rotated. The rock boring piddock has spines on the shell for boring. Exoskeletons may also be important camouflage devices and bear processes

The piddock, a bivalve that burrows in rock, exposed in its tunnel. The piddock uses cutting plates on its shell to bore through rock. The tough muscular foot seen in the bottom left of the picture anchors the piddock to the end of the burrow whilst the shell muscles open and close causing the shell to act as a rasp.

to break up the animal's outline or mimic some plant form. They may confer protective coloration or give gorgeous physical colours like those of metallic green Chafer beetles or blue-purple *Morpho* butterflies. The prickly spines of Sea urchins and the fine irritant bristles of some hairy caterpillars are defensive. The hairs on aquatic insects and spiders may be used to trap air bubbles allowing these animals to breathe underwater.

Skeletal Materials. Most animal skeletons are basically made of proteins such as collagen which have been rendered hard by the incorporation of mineral salts or toughened by chemical processes akin to the vulcanization of rubber and tanning of leather. The 'vulcanization' process links proteins into a firm mat by the introduction of sulphur; our hair, skin and nails are produced by a similar process, termed keratinization, and some invertebrate skeletons such as roundworm cuticles also contain keratins. In tanning, the protein fibres are linked together by quinone bonds to produce a dark brown toughened substance called sclerotin. Most insects, arachnids and myriapods owe their dark brown colour to the fact that their cuticles consist of tanned sclerotin. Proteins stabilized by sulphur or quinone are not the only substances to form skeletons. Amino sugars can be linked together to form the giant molecule or polymer, chitin which, though flexible, has great strength. Chitin occurs with sclerotin in arthropod cuticles and the internal transparent skeleton of squids,

called a sea pen because of its resemblance to a quill, is almost pure chitin. Many invertebrates have skeletons composed of several of these substances. Nematode worms have basically collagenous cuticles with both sclerotized and keratinized regions; crustaceans have protein, chitin and calcium salts in their cuticles. Mineralized skeletons are usually impregnated with either silicon or calcium salts.

Not all animals secrete their own skeletons. Some, like the shelled amoeba *Difflugia*, cement sand grains together to form a shell. The Caddis fly larva and marine fanworms are not content with the protection afforded by their own cuticles and build tubes which are often composed of materials that also form their natural background, they therefore provide themselves with an almost perfect camouflage. Fanworms use feeding currents to collect fine sand and mud particles which are cemented into a smooth tube with mucus, Caddis fly larvae use sticks, stones, small empty snail shells or even weave tubes of leaf fragments stuck together with silk. The soft bodied Hermit crabs use empty Sea snail and whelk shells as an adopted skeleton. There is even a nereid worm (ragworm) that lives with the Hermit crab for protection.

Coral Skeletons. Corals are similar to anemones except that each individual secretes a hard chalky skeleton. This skeleton is often white when the coral is dead but in life is clothed with a shimmering translucent film of living tissue. The true, or stony,

This hermit crab from off the Phillipines supplements its own skeleton not with snail shells, as is usual but with cockle and scallop shells.

corals may be either solitary or colonial. Solitary corals are rather like small Sea anemones only each sits in a cup of hard chalk. Colonial corals consist of a sheet of tissue containing internal digestive and canal systems. Here and there the tissue erupts into a mouth fringed with tentacles in groups of six. Underneath the polyp bodies is the skeleton. When the skeleton is cleaned, thin partitions radiating into the centre of each coral cup can be seen; these support the folds of tissue, called mesenteries, which divide the digestive cavity inside. The whole skeleton may be dome-shaped as in Brain corals, branching as in Stag's horn coral or form flat shelves. Other kinds of coral have tentacles arranged in groups of eight and are not reef builders, although some, like the Precious coral, may produce a hard skeleton. Coral reefs are formed in tropical and subtropical seas with mean annual

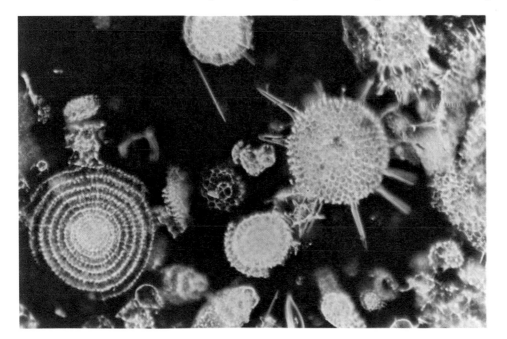

The silica-containing skeletons of dead radiolarian protozoans in radiolarian ooze dredged up from 1,460 ft (4,800 m) look like spun glass.

temperatures of 64–77°F (18–25°C) and corals are only able to grow in clear water at depths of up to 20–30 fathoms (120–180 ft, 37–55 m), although rarely less than 15 fathoms (90 ft, 27 m) down. Because of their requirement for shallow water, corals form fringing reefs closely bordering tropical shores or barrier reefs like the Great Barrier Reef, 1,200 mi (1,930 km) long off the north east coast of Australia. This is based on the continental shelf but is separated by a deep channel from the mainland. Atolls are ring-shaped coral islands enclosing a central lagoon which occur in mid-ocean, especially in the Pacific. They consist of rings of living corals overlying dead accumulated skeletons that may go down to depths of a mile. Darwin was the first person to wonder how corals, which need shallow water, can grow in mid-ocean and suggested that they were growing on the slopes of extinct under-water volcanoes which at one time formed a central island and gradually sank, leaving the lagoon. This theory has gained support from bore samples of the coral bed which turns out to be volcanic rock. But

why do corals need shallow water? The answer, it seems, is that they need light and the reason for this plant-like requirement is that coral tissues contain single celled algae (dinoflagellates) called zooxanthellae. These convert water and carbon dioxide excreted from the coral cells into sugars, but can only do this in the presence of light, a process called photosynthesis. Waste nitrates and phosphates produced by the polyps are also used by the algae to make proteins. The algae die in prolonged darkness. Coral polyps are carnivores and eat small animals captured by the stinging cells on the tentacles; if starved they do not eat their algae as a kind of side salad but cast them out. So what benefit is the coral gaining from their presence? It seems that the algae help corals to form their skeletons. In 1960 T. F. Goreau, working in the West Indies, measured the deposition rate of coral skeletons using radioactive calcium and found that growth was most rapid in bright sunlight, that a cloudy day might reduce deposition by 50 per cent and that hardly any growth occurred at night. The

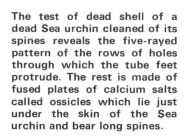

The test of dead shell of a dead Sea urchin cleaned of its spines reveals the five-rayed pattern of the rows of holes through which the tube feet protrude. The rest is made of fused plates of calcium salts called ossicles which lie just under the skin of the Sea urchin and bear long spines.

algae seem to aid the deposition of calcium taken up from sea water by the living coral tissues by absorbing excess carbon dioxide and bicarbonate ions from the system, for photosynthesis, leaving just enough for the optimum rate of combination of calcium and bicarbonate to form the skeleton of chalk, calcium carbonate. It is also possible that they may produce growth substances or provide optimum conditions for growth by removing waste products. This kind of relationship where two different kinds of organism are associated for mutual benefit is called mutualism.

Skeletons used for Floating and Drifting. Not all calcareous skeletons tie their owners to one place or slow down their pace. The jet set amongst molluscs, the cephalopods which use their mantle cavity and siphon for jet propulsion, have lightened their shells or lost them altogether as in the octopus. Many cephalopods that have retained the shell have adapted this as a buoyancy device by sealing gas into its chambers. The Pearly nautilus has a heavy chambered shell built in a flat spiral. It is beautifully tiger marked outside with cream bars on a tawny background but inside is lined with pearly nacre.

The 'slate pencil' Sea urchin *Heterocentrotus mammillaris* with its felt of short spines and stout primary spines that can be used as writing utensils. The spines, surprisingly, are large, single, crystals of calcium and magnesium salts.

A coral atoll formed in mid-ocean from the accumulated dead skeletons of millions of coral polyps. The coral can only grow in shallow water and the foundations of the atoll up to 1 mile under water are on the slopes of an extinct volcano.

A number of radiating curved partitions or septae divide the shell internally into chambers and the animal, which looks like an octopus but with about 200 tentacles and a head shield to seal the shell opening, lives in the last (most external) chamber. A full grown *Nautilus* has about 30 chambers and each is secreted in turn as the animal grows. Originally each chamber is filled with fluid, but as the nautilus strengthens the partition dividing the new chamber from the rest of the shell, some fluid is pumped out and a gas space formed. The nautilus grows very quickly and takes only about a year to reach full size. All the chambers are connected by a living strand of tissue called the siphuncle which contains an artery and a vein. This passes through a small calcareous spout facing inwards in each septum and these can be seen in empty sectioned shells. The gas inside the shell is mainly nitrogen and argon and is at a pressure of about 0·9 atmospheres. This lightens the heavy protective shell and gives buoyancy.

Nautilus lives in the Pacific at considerable depths of over 650 ft (200 m) where the pressure of sea water is very great, and its thick shell helps to withstand the crushing action of the sea at these depths. It can however migrate up and down in the water, and can change its buoyancy to do this by pumping liquid into or out of the chambers, thus decreasing or increasing the gas space. The fluid is pumped over the living wall of the siphuncle into or out of the blood system. The short calcareous tube around the siphuncle traversing each septum acts as a wick drawing fluid in contact with the water-transporting siphuncular membrane. The extinct ammonites had a siphuncle and could probably float in the same way.

The cuttlefish also uses its shell as a flotation device but this is comparatively reduced and is situated internally under the skin of the back. Cuttlebone is commonly seen washed up on the seashore and is often given to cage birds. It is a shield-shaped structure almost as long as the cuttle's

Many different kinds of coral occur on the Great Barrier Reef.

body with a horny margin and chalky centre honey-combed with fluid filled chambers. A membrane under the shell pumps fluid into and out of the shell in the living animal to alter the density for swimming or sinking. There is some evidence that this process is diurnal and corresponds to the activity phases of the cuttles, allowing them to be light and buoyant by day when feeding on shrimps and crabs and to sink to the bottom at night.

Other animals that use their skeletons as a buoyancy device are the By-the-wind-sailors and their allies. These look like jellyfish but are actually a colony of polyps hanging from a flat, horny, gas-filled float which has a projecting sail on its surface. They are common in warm waters and have the sail at an angle to the axis of the body so that the animal can be driven along by the wind. The sail is diagonal and may run either from right to left or from left to right across the float and it is said that offspring of any one parent show variation in this respect, a proportion having the sail running one way and a proportion having it running the other. This may allow the young to drift in different directions from the parent. Different kinds of buoyancy devices are employed by planktonic crustaceans floating in the world's oceans. These often have flattened bodies with long feathery spines on the legs and tail and the zoea larva of the Porcelain crab has an enormously long unicorn-like spine extending forwards from the head. The planktonic crustaceans of tropical waters need even more spines than those in colder waters because warm waters have a lower viscosity than cold waters, so sinking tends to be more rapid. Non-skeletal aids to buoyancy in animals include accumulation of fat or low density ions like ammonium chloride. The ammonium ion is sufficiently lighter than sodium to give marine animals an upward thrust. Cranchid squids keep a pocket filled with ammonium chloride for buoyancy.

Growing and Moulting. Corals clothe their skeletons and can easily add to them, but how does an animal increase in size when its skeleton surrounds it? Animals with chambered shells like the radiolarians and nautiloids simply add new and larger chambers as they grow, sealing up the old ones behind. Snails produce new shell whorls with a progressively increasing diameter and confine their soft deformable digestive gland to the small inner whorls.

Living coral, a hard inner skeleton covered with jelly-like tissue, and soft polyps—the so-called coral animals.

Bivalve molluscs start with two small lobes of shell and as the mantle (skin) grows around the edges, a rim of new shell is secreted producing the characteristic growth lines. Segmented worms have cuticles made of a lattice work of collagen fibres which presumably stretch as the worm grows. Roundworms and arthropods, however, have to moult at intervals. Roundworms moult four times during their lives and arthropods moult many times. The insects do not usually moult as adults although some of the more primitive forms like silverfishes do. When an insect moults special skin glands produce secretions that digest away the inner layer of the old cuticle so that a space develops between this and the new cuticle that is forming underneath. When the new cuticle is ready the old skin breaks and out steps a newly clad insect in a cuticle that is at first wrinkled, soft and colourless.

Section through the body wall of a coral to show the algae called zooxanthellae (1) enclosed in the coral's tissues. The presence of these algae explains why corals require light and are therefore restricted to life in shallow water.

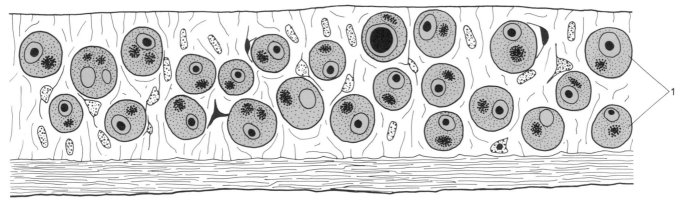

A By-the-wind sailor showing the flat, horny, gas-filled float from which a colony of polyps are suspended. The float has a triangular sail that propels this strange creature across the surface waters of the ocean.

This quickly smooths out as the insect pumps air into the body and it tans to a leathery brown in the presence of air. The moulting process is under hormonal control but may be initially triggered by a feeding response. In the blood sucking bug *Rhodnius* each juvenile stage has one blood meal. This distends the abdomen and stretch receptors in the walls send messages to the brain which cause certain brain cells to produce a neurosecretory substance. This is carried in the blood to the thoracic glands and causes them in turn to secrete a hormone called ecdysone (which has been isolated from silk moth larvae in crystalline form). Ecdysone stimulates the epidermal cells to grow and divide and to secrete the new cuticle. Sometimes animals that moult retain the uncast old cuticle as an outer protective sheath. This occurs in the formation of the pupal stage of many insects such as butterflies and is common practice in parasitic nematodes where retention of uncast cuticles often makes them very resistant to drying.

Our Hermit crab with its soft body does not

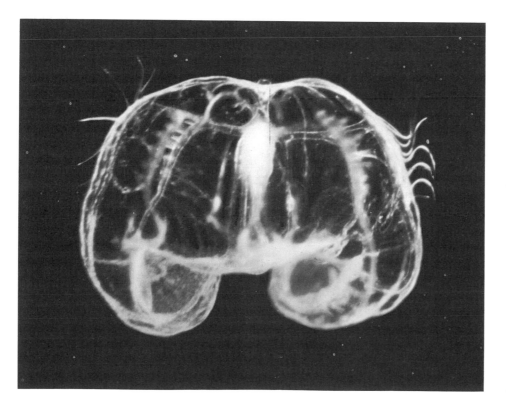

The young Sea gooseberry moves by means of comb plates made of fused cilia visible at right.

The cleaned skeleton of the Venus's flower basket sponge is made of glistening silica spicules woven into an intricate design.

escape the hazards of moulting altogether. As the young Hermit crab grows it has to select larger and larger shells to occupy. As a rule only young Hermit crabs are found on the shore and these live in small Top shells or Dogwhelk shells. Larger Hermit crabs occur offshore and occupy large whelk shells. A certain insecurity is suggested by the way that hermit crabs associate with stinging anemones sometimes plucking them off the sea bed and planting them on their shells for protection. One species of Hermit crab retains a shell that becomes too small and covers the gap with the anemone *Adamsia* which wraps its foot completely around the crab. This relationship is very specific for *Adamsia* is never found without this crab species.

Ciliary, Flagellar and Amoeboid Movement. Not all animals use muscles to move. Small animals under about 0·24 in (6 mm) in length may use the beating of many minute lash-like processes called flagella or cilia to propel themselves. These are mainly of use in water and are not in general found on the outsides of terrestrial animals. There are, however, always exceptions and some large terrestrial flatworms, although they have muscles, use cilia to glide on a mucus slime secreted from the body. Snails too may use cilia and mucus as well as muscular waves for gliding along. Many single celled protozoans are propelled by cilia and flagella and the microscopic transparent marine larvae of common sea shore animals like clams, Sea stars and segmented worms move by means of bands of cilia. Sea gooseberries move by the beating of iridescent combs of fused cilia arranged longitudinally down the body. It is difficult to distinguish absolutely between cilia and flagella but in general cilia are short and numerous and beat with a straight down stroke and bent recovery stroke, whilst flagella occur usually singly or in small numbers, are long and have several waves passing along them at once. These waves may all lie in the same plane or may have a corkscrew twist as they travel along the flagellum from base to tip. Most animal sperms have a long

23

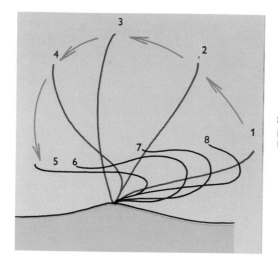

Stages in the beating of a typical cilium, the lash like processes used to power many single celled animals. Forward beat numbered 1-4 in red, beat recovery stroke shown in black, numbered 5-8.

flagellate tail that pushes the sperm head along by a kind of rippling movement rather like an eel pushing a balloon in front of it! Flagellate protozoans that produce spiral waves in the flagellum are actually pulled around by this so that the whole body gyrates along a spiral path as they swim. Ciliates such as slipper shaped *Paramecium* and *Spirostomum* are covered with cilia which beat in waves, usually compared to the waves produced on a wheatfield by wind. The cilia obviously work in a co-ordinated way but how this co-ordination is produced is not really known. Some biologists noting that the cilia are sometimes joined by fibres or kinetodesmata suggest that these co-ordinate their activities, others favour waves of electrical excitation passing over the cell surface as the co-ordinating system. Comparatively little, too, is known about what causes the flagellum or cilium to bend. The internal structure of cilia and flagella is very similar, both having an internal framework of minute tubules (microtubules). These have a remarkably constant arrangement throughout the animal kingdom so the cilia of *Paramecium*, the flagella of sponge cells and the cilia lining the bronchial passages of man are almost identical! There is always an outer ring of 9 double tubules each with a pair of arms and an inner pair of two separate tubules. Studies on mutant flagella have shown that unless this 9+2 structure is present the flagellum cannot beat, which suggests that microtubules play an active part in contraction. One theory suggests that they slide past one another rather like muscle filaments.

Amoeboid Movement. Amoeboid movement by cytoplasmic streaming may seem simple but in fact this is a very specialized process and different kinds of amoebae, for instance, produce different shaped pseudopodia, some being long and thin others

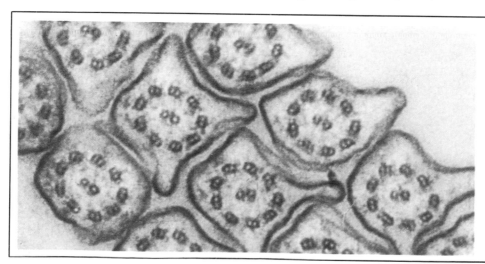

Electron micrograph showing the 9+2 arrangement of hollow microtubules in a number of closely packed cilia which have been cut across.

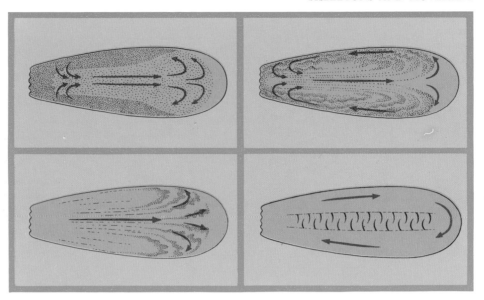

Diagram of *Amoeba* giving three of the theories which explain how cytoplamic movement might occur. One explanation is that fluid cytoplasm called sol is pushed forward like a fountain from behind and forms firmer gel (dark stipple) as it reaches the edges of the animal (top left and right). Bottom left illustrates an alternative explanation that cytoplasm at the front end contracts pulling the *Amoeba* forwards. Bottom right shows an explanation based on their being a ratchet system that rotates the animal forward.

blunt and lobose. Heliozoans or 'sun animals' are amoebae that produce long, fine semi-permanent pseudopodia supported inside by spiral rods of micro tubules. The animals roll along on these spiky processes which pick up food at the same time. No completely satisfactory explanation is available for the way *Amoeba* pushes its cytoplasm around. One theory suggests that the cytoplasm can exist in two interconvertible states, a firm gel and a fluid sol. During movement it is thought that the sol flows forwards then becomes gel around the edges to channel forward movement. Other theories suggest that *Amoeba* has very fine muscle-like filaments (microfilaments) that can contract to distort the cytoplasm, thus dragging the animal forwards.

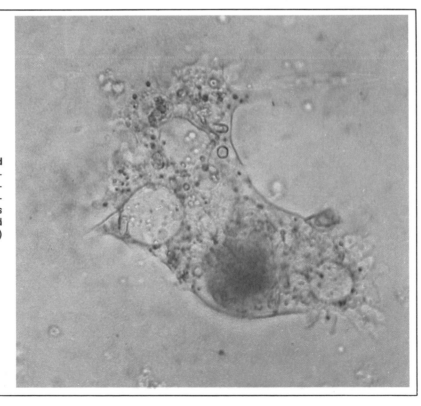

Amoeba, a single celled animal, viewed under the microscope. The greyish, granular body of this animal is made of cytoplasm which streams out to form pseudopodia, or false feet, as *Amoeba* moves forward. The nucleus (large dark spot) and contractile vacuoles (lighter circular areas) are clearly visible.

The European lugworm, a species of *Arenicola* removed from its sandy burrow. The lugworm's body is filled with fluid which acts as a hydraulic skeleton during burrowing. The bristle-like chaetae seen projecting in groups from each segment help the worm to grip the sides of its burrow. The dark red tufts are external gills.

Fluid Skeletons and Movement. The soft bodied jellyfishes, Sea anemones and worms are not the only forms to use fluid skeletons for locomotion. Shelled animals such as the Sea urchins and clams use their outer skeletons mainly for protection and have a fluid skeleton for movement. Sea anemones would collapse if they were not filled with water and in fact pump water into the gut cavity which extends into the tentacles, giving themselves a firm core for the body wall muscles to act on. This fluid skeleton helps to stretch muscles back to their resting length after contraction because muscles cannot actively relax. Contraction in one part of the body causes pressure build up in the fluid which bulges out a neighbouring region, thus stretching the walls. The fluid skeleton is mainly used in extending the tentacles and in bending of the body. Anemones can also shuffle along on the basal disc and some anemones like *Edwardsia* can burrow in sand. Jellyfish and flatworms use the turgor of their tissues to give their bodies firmness and roundworms, which have fluid under high pressure in the body cavity, use this as a kind of stiff rod for the longitudinal muscles to pull against as they swim. The segmented worms like the earthworm have a series of separate fluid filled compartments, each surrounded by two sets of muscles, one running round the body and one running longitudinally. As one set of muscles contracts the fluid pressure increases in a direction which aids the other set of muscles to extend. The earthworm burrows by rhythmic move-ments caused by alternate contractions of the circular and longitudinal muscles. When the circular muscles contract the body becomes long and thin and is pushed forwards and as the longitudinal muscles contract the body shortens and becomes fatter. As waves of contraction pass down the body towards the tail, each segment alternately elongates then contracts, while the earthworm wedges itself against the burrow by protruding bristles called chaetae. More than one wave of contraction may pass down the body at once so that several regions may be in contact with the walls of the burrow at one time. Marine burrowers like the lugworm use their fluid skeletons in a different way. They have lost the septae forming fluid tight bulkheads between each segment and have a more or less open fluid system except in the head and tail regions. These worms push all the body fluid to the front by alternate contractions of the body wall muscles and cause the head to be forcefully thrust into the sand. Clams, Razor shells and other bivalves use a hydraulic system for burrowing. They have a long slender muscular foot which is protruded between the shell valves, extended and inserted into the sand, then pumped full of blood. The swollen foot is thus anchored into the sand and the shell retractor muscles contract, pulling the body down on to the foot. The cycle is repeated after deflation of the foot. The Sea stars, Sea urchins and Sea cucumbers have a very special fluid system that hydraulically operates a number of tube feet, each with a bulb region

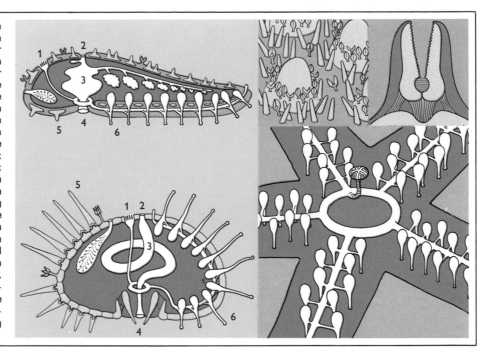

The Sea star and Sea urchin are much more similar than might be supposed from their external appearance. The Sea urchin is really like a Sea star with its 5 arms arched over the central disc and fused. Above left shows a section through the disc and one arm of a Sea star. Below left, a section through a Sea urchin: (1) sieve plate; (2) anus; (3) intestine; (4) mouth; (5) calcareous spine; (6) tube feet. Above right, a portion of the surface of a Sea star showing spines projecting from hard ossides under the skin, and right, a pedicellaria or forceps-like structure which has a skeleton of its own and prevents organisms settling on the skin surface. Below right, part of the water-vascular system in a Sea star showing the bulbs of the tube feet connected to a central canal along each arm.

inside the arm connected to an extensible tubular portion bearing a termal sucker. In a Sea star the two rows of tube feet on each side of the arm are connected to a central vessel and this in turn joins to a ring canal in the central part of the body. A further canal runs to the top of the body where it opens on a stony plate containing pores. Protrusion of these armies of feet is effected by the bulbs at the top contracting and squeezing fluid into the tubes, causing them to extend and attach by means of the sucker. They are stiff enough to act like levers pushing the arm along. Withdrawal of the tube foot is effected by contraction of longitudinal muscles which join each foot to a collar of hard plates. Sea

The underside of one of the arms of the European Common sea star *Asterias rubens* showing the two rows of hydraulically operated tube feet. The tube feet are used as levers for walking and have small suckers at their ends which give a firm grip on slippery surfaces. They are also used by the Sea star when it opens oysters and mussels.

The edible Sea urchin *Echinus esculentus* of European coasts climbing the glass wall of an aquarium with its long tube feet. The tube feet emerge from the hard spiny test, or shell, along the five pairs of pink grooves. The mouth which is lined with a powerful circle of hard teeth used for scraping algae off rocks, is seen in the centre of the animal.

stars open oysters and mussels using their tube feet in conjunction with body muscles and, by wrapping their arms around the shell, they can exert a pull of 12 lb (5·4 kg) for periods as long as six hours. Once the shellfish gives way the Sea star pushes its stomach between the valves and starts to digest the soft insides, taking no note of the correct way of eating oysters by swallowing them whole.

Walking and the Arthropods. The tubular skeleton of arthropods gives great strength coupled with lightness. A tube requires more force to break it by bending and is less likely to buckle than a solid rod of the same length made of the same amount of material. The evolution of a jointed tubular limb system has allowed the appendages to perform very finely controlled movements and in insects and arachnids has allowed the body to be picked off the ground for faster movement. The centipedes have achieved faster movement, partially by picking the body off the ground but also by increasing the length of the legs so that small muscle contractions are translated into long strides. The insects have reduced the number of legs to six and use faster movements of each leg rather than having long legs to gain speed. Insects move their legs in a different order depending on how fast they are moving.

The mechanism of jumping in springtails: (1) resting position with muscles relaxed and low fluid pressure in abdomen and furca; (2) abdomen wall muscles contract, causing high fluid pressure in furca; (3) simultaneously, muscle to 'teeth' contracts, releasing furca, which flicks away from body, propelling springtail forwards (4).

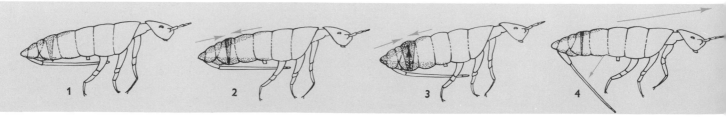

The House centipede *Scutigera forceps* is the fastest moving of the centipedes and can reach speeds of 20 in (50 cm) per sec in order to catch insects. Its high speed is partly due to the length of its legs.

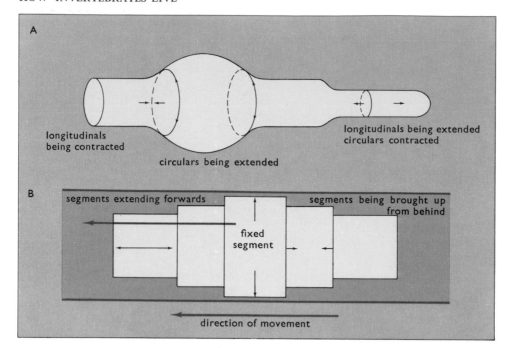

A
longitudinals being contracted
circulars being extended
longitudinals being extended circulars contracted

B
segments extending forwards
fixed segment
segments being brought up from behind
direction of movement

Shape distortion and movement in a worm provided with circular and longitudinal muscles acting on a fluid-filled cavity (hydrostatic skeleton) of fixed volume: (A) changes in shape; (B) movement.

When they are moving very fast they tend to move two legs from one side and one from the other at once, leaving a tripod in contact with the ground, whilst the other legs are moving forwards.

Jumping and Pole Vaulting. Some joint legged animals also use fluid for jumping. The Jumping spiders are small spiders less than ½ in (12·7 mm) long that don't make webs but pounce on their prey. They use their back legs for jumping and instead of just using muscle power to straighten the legs they pump blood into them, causing them to extend rapidly, pushing the body off the ground.

Large brown jellyfish swimming by pulsations of its bell, which expands and contracts like an umbrella being opened and closed. This is a form of jet propulsion.

The squid swims backwards with the pointed end first and tentacles trailing behind. The thrust comes from a form of jet propulsion as water is expelled from a moveable muscular siphon held under the body.

Jumping spiders spin a line of silk as they jump and providing this is attached at one end it prevents them from somersaulting in the air. Some small soil insects known as springtails or Collembola use a kind of pole vault leap to progress. They have a stiff rod or furca formed from a modified pair of primitive legs situated right at the end of the abdomen. This is usually kept clipped in position under the abdomen, but when the abdominal muscles contract rapidly blood is forced into the lower abdomen and furca, the clips holding the furca are simultaneously released and the furca shoots downwards and backwards, striking the ground and lifting the animal forwards. The flea, of course, is the jumper par excellence and distances for record jumps are as much as 12 in (31 cm). Propelled by its muscular back legs the flea sails through the air, turning over perhaps several times to land facing the way from which it came. The first two pairs of legs act as grappling irons and are held outstretched during the jump on to the "hairy windswept cliff" of the host, as Miriam Rothschild puts it. To power the jump, the flea uses the elastic snapback stored in two pads of elastic material called resilin at the top of the back legs when the leg tenses up before the jump.

Flying and Gliding. The power of sustained flight has been achieved only four times in the animal kingdom in birds, bats, the extinct pterodactyls and insects. The insects primitively have two pairs of wings on the thorax like dragonflies and these are flat plates of thin cuticle containing a network of air-filled tubes (the 'veins'). The wings of insects are attached to the thorax by a complex joint that enables them to be tilted or rotated in flight so that lift is obtained on both the upstroke and down-stroke. During the downstroke the leading edge is down whilst the back edge is raised; on the upstroke the direction of wing tilt is reversed. Dragonflies use both pairs of wings to fly but there has been a tendency for insects either to link front and back wings together by bristles so that they act as one unit, as in bees and butterflies, or to use only one pair of wings for flight. Beetles have converted the first pair of wings into hard wing cases but in the flies and mosquitoes the front pair of wings is used in flight and the back pair have become reduced to stumps called halteres which are balancing organs. Butterflies with scale covered wings have a lazy flight with a wing beat frequency of about 9 beats per second, houseflies beat their wings 180–200 times per second and mosquitoes 1,020 beats per

31

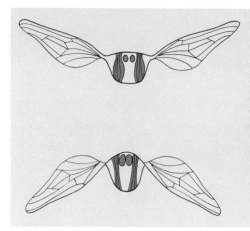

Movement of the wings of insects works upon a principle quite different from that of birds. The flight muscles are not attached to the wings but to the body wall. When the vertical muscles contract, the top surface of cuticle flattens and the wings snap upwards; when the longitudinal muscles contract the top surface of cuticle bulges outwards and the wings snap downwards.

second. These high speeds in flies and mosquitoes are obtained by the wing muscles exploiting the elasticity of the thoracic cuticle. The muscles are attached not directly to the wings but to the cuticle. This contains the elastic material resilin at the wing hinge. As either set of wing muscles contracts strain builds up in the cuticle, the wing is forced to move through a midpoint and then is suddenly flicked into either the top or bottom position as the strained cuticle snaps back. The opposing set of muscles then contracts, deforming the cuticle in another direction for the recovery stroke which is again terminated by shoot-through of the wing as the elastic energy of the cuticle is released. This device of using the elasticity of skeletal structures to complete a movement is known as a click mechanism. It is mechanically most efficient since the wing muscles can work at almost constant length and only trigger the change in direction of the wing stroke. It also overcomes the problem of having nerves that will fire fast enough to control muscle contraction. In the case of this click mechanism the system oscillates until the next nerve impulse comes

along. The fast wingbeats of mosquitoes achieved by click mechanisms have an offbeat function. The whine of the mosquito wings is used as a mating call. The females have a louder and deeper call than the males, which burst into an answering chorus when they hear the females. Worker bees use their wing beats not only in flight and as auditory signals but to cool the hive by fanning air through it. Many insects, especially butterflies, use wing patterns and movements as visual signals.

Flying squid can launch themselves into the air by a jet propelled thrust that carries them for 10–20 ft (3–6 m) out of the water for distances of over 50 ft (15·2 m). All squids use jet propulsion and swim backwards with the pointed streamlined end first when moving fast. The thrust comes from the muscular water filled pouch (mantle) under the body that opens by a funnel. Water enters around a collar region behind the head and this opening is then closed as water is forcibly expelled by contraction of the body wall muscles through the funnel. The funnel can be swivelled to face forwards or backwards so that the squid can advance or retreat.

Foodchains, Eating and Being Eaten

All animals ultimately depend on plants for their food because animals need to consume complex food stuffs which do not in general occur in the environment but are present only in the bodies of other living organisms. Plants are the producers in the ecosystem because they can manufacture complex food molecules from simple inorganic substances like carbon dioxide, water, nitrates and phosphates, using the energy of sunlight to do so. Animals obtain food substances by eating plants directly (herbivores), by feeding on the bacteria and fungi that decompose plant remains or by eating other animals (carnivores). This leads to the establishment of food chains or food webs in which plants are eaten by several kinds of herbivores, the herbivores are eaten by a variety of carnivores and carnivores may be preyed upon by, usually, larger carnivores. Animals are not always either herbivores or carnivores, some eat both plant and animal food. Scavengers and parasites also enter into the web of feeding associations and in particular are often highly specialized to certain diets. The role of bacteria and fungi as decomposers should not be underestimated because these break down dead protoplasm and release simple substances which are recycled for use by the producers. Food webs depend upon carefully balanced systems of populations and change in the numbers of even one species population in a food web may produce far reaching effects throughout an entire ecosystem.

Food chains and the conversion of one kind of food into another produce a phenomenon known as a pyramid of numbers with many producers at the bottom and decreasing numbers at each succeeding level. Because at the bottom of the food chain the plants have to support all the animals and decomposers that depend on them they have to be extremely abundant. At the next level, herbivores eating plants have to produce sufficient offspring to satisfy the carnivores that depend on them. At succeeding levels, size becomes important and the numbers of large animals have to be limited so that the numbers of organisms available to them as food can support their bulk. Energy conversion from one level to another is in fact very inefficient. To increase in weight by one ton a Baleen whale has to eat 10 tons of small shrimps known as krill. Each ton of krill has been nourished by 10 tons of phytoplankton so this means that each ton of whale is equivalent to 100 tons of phytoplankton. About 90 per cent of the food energy is lost at each stage of conversion and only 10 per cent is used to build up the body of the whale. The rest is used for movement, reproduction, temperature maintenance and other vital processes.

One of the secrets of successful eating is to specialize. Animals tend to become so highly adapted to finding, consuming and digesting a particular diet, that they outfeed their competitors, grow fat and multiply. In their search for a niche,

A pyramid of numbers shows that animals such as carnivores in the second trophic level are greatly outnumbered by herbivorous members of the first trophic level. During its life, a whale is responsible, directly or indirectly, for the consumption of many millions of phytoplanktonic individuals.

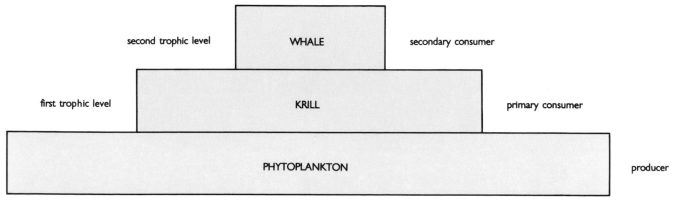

second trophic level	WHALE	secondary consumer
first trophic level	KRILL	primary consumer
	PHYTOPLANKTON	producer

invertebrates have exploited unlikely sources of food. The Marine shipworm, a modified bivalve and the gribble, a crustacean, both bore into and damage submerged wooden structures feeding on wood fragments and the fungi that decompose them. Termites, Timber and Bark beetles and Wood wasps also eat wood. The Clothes moth caterpillar is one of the few free living animals to tackle hair (wool) with relish since this is very difficult to digest, but the Dermestid beetles which infest drying carcasses in museums eat their way through skin and bone. Dung is perfectly acceptable to other beetles such as the Dor beetle burrowing into cowpats and sacred Scarab beetle which stores balls of dung underground. The Ancient Egyptians regarded this beetle as sacred and saw the balls of dung fashioned by them as symbols of the earth.

Many plant feeders are very specialized, sometimes confining their diet to a single plant species and perhaps eating only leaves, roots, bark, nectar or even pollen grains (as do some springtails and bees). Parasites of course are very highly specialized to particular diets and conditions. Fleas can survive on blood meals from alien hosts for a while but need a blood meal from their chief host before they can reproduce. Some Tsetse flies prefer to feed on game but will bite man as a second choice. One nematode worm feeding on onions eats only the cell wall but without releasing the cell cytoplasm which is poisonous to it.

Defences Against Being Eaten. Feeding is a much more complicated business than one tends to imagine in this, the age of the fish finger. It is complicated to a large extent by the fact that many animals and indeed plants actually try to avoid being eaten and have effective behavioural or chemical defence methods rendering them unavailable, invisible or unpalatable. Some plants like thistles and cacti have defensive spines, nettles have a sting and in fact many plants produce irritants and poisons to deter their predators. It is well known that foxgloves and milkweeds contain heart stimulants (digitonin), perhaps less well known are the facts that potato roots and shoots contain the poisonous glycoside solanine, that buttercups can release a strong irritant, protoanemonin, or that St. John's Wort *Hypericum* secretes a crimson dianthrone (hypericin) that causes skin irritation and can produce blindness and wasting if eaten. Plant alkaloids such as tubocurare, used on blow pipe tips by South American Indians, morphine, ephedrine mescaline, lysergic acid and nicotine have powerful pharmacological effects on nervous tissue. Nicotine can cause paralysis of aphids feeding on tobacco plants. Several plant species also produce protective substances that mimic the effects of insect hormones and when eaten by insects disrupt normal metabolism. Pine trees which are subject to insects attacking the trunk produce so-called phytoecdysones which stimulate insect moulting.

The Sea hare is a slug-like marine mollusc which browses on sea lettuce and other seaweeds. As it lacks a protective shell it relies on camouflage to protect it against predators and besides having pigment spots on its skin uses the pigments from the plants it feeds on to help it match its weedy background.

Two examples of marine food webs; the top one starts with a living plant (seaweed), the bottom one with dead seaweed. The living seaweed, or more usually single celled algae (phytoplankton—not shown) are eaten by herbiverous animals such as small shrimps, molluscs or in the case of the single celled algae by filter-feeding larvae, Sea squirts and other animals. These herbivores then form the prey for larger animals which are carnivores and these in turn are eaten by still larger carnivores and so on. Man is very often at the ends of food chains because nothing tends to eat man! The bottom food web involves bacteria which are very important in decomposing dead material. Several kinds of filter-feeding invertebrates feed on bacteria by sieving them out of the water. As much energy is lost at each level of the food web it would be more economical if man were to eat dead seaweed (or phytoplankton) directly, but fortunately the world food problem has not yet become so acute that this is necessary.

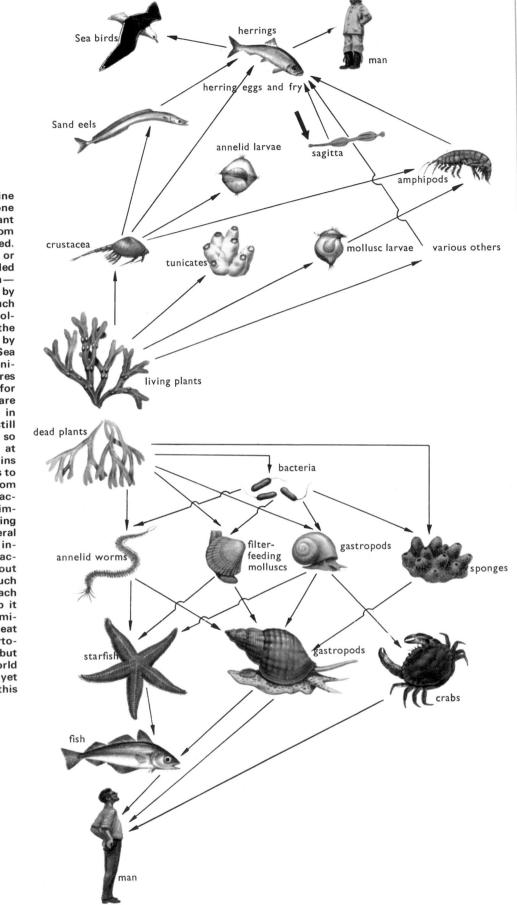

Above a critical dose these produce fatally accelerated development. Juvenile hormone analogs occur in the Balsam fir and other plants and arrest development at an immature stage in insects consuming them. In the sea phytoplankton can produce toxins and the deadly red tides caused by blooms of poisonous marine dinoflagellates that may kill all life for miles are well known. In fresh water dense ageing bodies of the alga *Chlorella* produce a secretion said to slow the filtering rate of the Water flea *Daphnia* that feeds on it. This may produce some kind of control on the grazing of the crustacean. Inevitably particular animals here and there have managed to counter these subtle defences evolved by plants. Beetles of the genus *Chrysolina* manage to feed on St. John's Wort, having evolved a mechanism which counteracts the effects of hypericin, and thus exploit a food source untouched by other herbivores. One species of *Chrysolina* actually uses the repellent as a cue to locating the food plant. The Cabbage white butterfly too uses the supposedly irritant, mustard oils present in cruciferous plants to locate its food. Pyrethrins, used commercially in insecticide dusts, are present in various kinds of African daisy but there is some evidence that moths and butterflies feeding on these and other plants have enzymes which render these natural insecticides harmless.

Animal defences against being eaten are more widely recognized than the subtle chemical deterrants used by plants. Fast movement is important for flight or withdrawal from potential predators. Animals may also use either warning postures and patterns to alarm or deter predators or camouflage to hide themselves. Eyespot patterns, for instance, are commonly used in insects to deter mainly bird predators. These are round spots often with a contrasting ring or rings which occur on the sides of the body of some caterpillars and on the hind wings of various butterflies. In many cases the eyespot is made conspicuous to the would-be predator by a sudden change in posture. Many inedible or venomous animals are brightly coloured and dazzlingly patterned as a warning to would-be predators. The yellow striped wasps and hornets provide an example of this. The evasive tactic of camouflage is used by many invertebrates. The drifting pelagic jellyfishes, crustaceans, bristleworms, molluscs and tunicates are transparent, which makes them very difficult to see. Similarly many shore-living snails, crabs and worms match the background of sea weed and stones amongst which they live and some change colour according to the kind of algal diet they are presently eating, thus becoming very inconspicuous amongst the fronds of their food plant. The Sea hare, *Aplysia punctata*, does this, changing from dark red to green as it changes from a diet of red to green sea weed. Shrimps, prawns and other crustaceans can change colour to match their background, an event controlled by hormones secreted in the eyestalk. Camouflage is not only used as a defence, it is also

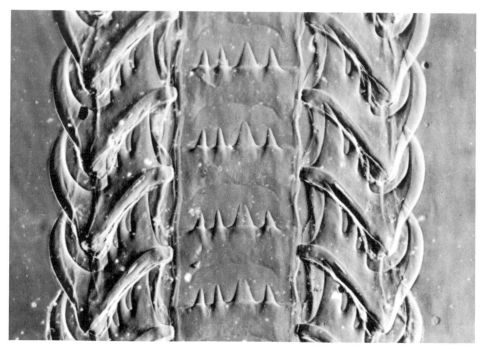

The rasp-like tongue or radula of a whelk bears many rows of teeth and is continuously produced throughout life as it is worn away in use. These file-like tongues were probably originally adapted for scraping algae films off rocks and browsing off plant material.

Eyespot patterns on the hind wings of the moth *Automeris io* are flashed at would-be bird predators to startle them.

used by predators enabling them to make surprise attacks. Dragonfly nymphs, for example, are brownish in colour and lie very still amongst weeds. Concealed in this way they prey on other aquatic insects, fish fry or even tadpoles by shooting out their modified lower lip which bears hooks for grabbing the prey.

Chemical defences used by animals against predation range from the inks secreted by octopuses and related forms and the stings of jellyfishes and Sea anemones to insect bites, stings and stinks. Bombadier beetles have a most effective explosive deterrent and spray a blast of hot bitter liquid from the abdomen when molested. The spray is produced by combination of two secretions, one containing quinones the other hydrogen peroxide. These substances are mixed with an enzyme in a special hard walled reaction chamber and oxygen released from the hydrogen peroxide provides the propellent for the spray. This noxious blast is enough to deter even large predators such as toads. The North American Darkling beetle is another form that produces a defensive secretion and, when disturbed, stands on its head and sprays the attacker from glands in the abdomen. An almost identical beetle, *Megasida*, assumes the same posture when attacked but contains no defensive secretion! One very specialized defence reaction to a predator occurs in the microscopic freshwater rotifers. It seems that

a predatory rotifer *Asplanchna* releases into the water a substance that causes the eggs of another rotifer *Brachionus* to develop into individuals with long moveable spines that are otherwise absent. These spines give some protection against being eaten because they make *Brachionus* more difficult to ingest.

Escape Responses of Marine Invertebrates. Some apparently sluggish marine animals can become suddenly very lively if not mildly hysterical when confronted with certain predators, and make remarkably rapid responses that help them to escape. Many invertebrates such as limpets, snails, clams, scallops and Sea urchins are particularly wary of Sea stars and some will produce their specific escape behaviour when experimentally exposed to extracts of Sea stars. Cockles and a few species of clams make a very dramatic exit when touched by a Sea star. The bivalve opens its shell, thrusts the foot out against the ground it is resting on and gives a violent push that sends it hurtling 4–5 in (10–12·5 cm) away. Scallops are much more active bivalves and are able to swim off or leap up and down by opening and closing the shell valves, like a pair of flapping false teeth, if confronted with a predator. Whelks *Buccinum undatum* also resent the many footfalls of an approaching Sea star and throw their whole body into violent rolling motion, first in one direction, then in another, literally

37

Slugs are herbivores and can do much damage to garden plants. They have a special long tooth-covered tongue which rasps up hard plant tissues. The breathing part of one of the slugs is visible.

shaking off the attacker. The fright reaction of abalones *Haliotis* to Sea stars has been appreciated by the Maoris of New Zealand for some time. When touched by a Sea star the abalone raises its shell on its foot and twirls it vigorously in the water. The Maoris, who like many other nations prize the abalone as a great delicacy, fish for this taking advantage of the escape reaction. The fishermen touch the abalone with a Sea star and as the mollusc lifts its shell, pluck it from the rock. Normally, like the limpets to which they are related, the abalone is very difficult to detach. The abalone may produce an alarm substance that causes other abalones in the vicinity to produce an escape reaction too. Even some Sea anemones seem to distrust Sea stars and rapidly lose hold and creep or swim away. Sea cucumbers show unsuspected natatory talents when confronted with their unfriendly and very distant relations, the Sea stars. This group also practises evisceration. They simply split the body wall and eject the gut and respiratory organs to save their skin (this being all that is left). The predator is left either confused or busy eating the unexpected benison before it. Some Sea cucumbers prefer to keep most of their insides and have special organs at the back end which are exuded when the animal

The ormer or Ear shell is related to the limpets and browses algae off rocks. There are a series of holes along the shell through which a respiratory current is drawn. The ormer literally shakes off attackers by rolling its shell from side to side.

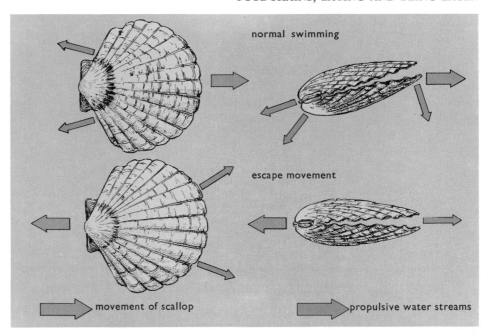

normal swimming

escape movement

movement of scallop

propulsive water streams

The scallop uses two quite different swimming actions, the first for rapid swimming and the second for rapid escape. The direction in which the propulsive water currents emerge is controlled by the flaps of mantle lining the shell.

is threatened. These form very sticky threads and can often trap an intending predator. The British Sea cucumber *Holothuria forskali* is sometimes called the cotton spinner because of the ability to discharge protective threads. Keyhole limpets, which are like ordinary limpets but with a hole at the top of the shell, have a neat device to repel boarders. They send a slippery sheet of mantle wall creeping up the outside of their shells. The Sea star is unable to grip onto this and falls off. Sea urchins actually nip intruders with their poisonous pedicellariae. Some are also sensitive to the shadow of a predator, this causing the spines to move comparatively rapidly, which might discomfort an invader or be part of an escape reaction. Scallops, cockles and tubeworms are also amongst the marine animals that use shadows as an early warning system of impending attack. The scallop has a special kind of eye for responding to shadows, as will be mentioned later.

Plant Feeders. Animals feeding on plants, or plant materials such as wood or on fungi or bacteria need special adaptations for their diet. Plants and fungi have strong cell walls which have to be broken down by digestive juices containing cellulase enzymes and often by mechanical action involving jaws and gizzards. Plant feeders such as roundworms and aphids have long hollow stylets which are inserted through the cell wall and used to suck out the cytoplasm. Some roundworms have solid spears used like battering rams to burst a hole in the cell walls. Often plant feeding animals have no

The Sea slug *Elysia viridis* matches perfectly the green weed *Codium* on which it crawls because it stores the chloroplasts of its food in its own tissues thereby becoming invisible to predators.

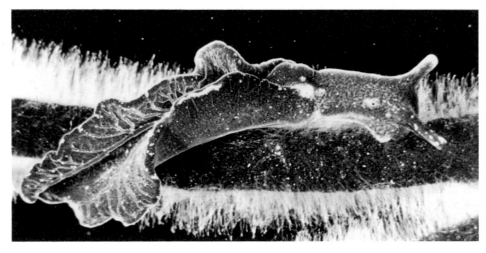

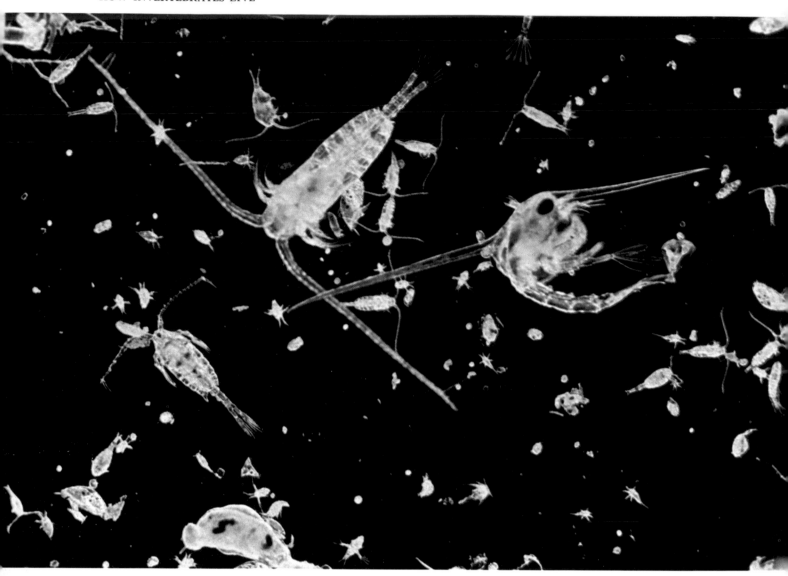

Marine zooplankton consists of a 'soup' of microscopic transparent animals that live on single celled diatoms and are the primary consumers in the sea. The two largest are *Calanus finmarchius* and the zoea larva of the masked crab which has enormous body spines to help keep it afloat.

cellulase enzymes of their own but use protozoan and bacterial symbionts in their gut to produce these. Termites have many flagellate protozoans in their gut which ingest and break down wood particles on which the insects feed and are then themselves digested by the termites. Because of the comparatively small amount of protein available in plant food compared with its overall bulk, plant feeders tend to eat continuously. Despite the defences used by plants mentioned earlier, there are animals specialized to eat many different kinds of plant. Invertebrates feeding on land plants include nematodes, snails, slugs and earthworms. The earthworms feed mainly on decaying plant material

and drag leaves into their burrows tip first. They presumably have a sense of taste since, as Darwin noted, they show a marked preference for certain kinds of leaves. It has been calculated that an average acre (0·4 hectare) of soil may contain some 3 million earthworms weighing about 15 cwt (760 kg) and that these move about 18 tons (18·25 tonnes) of soil a year. The soil mixing produced by earthworms is very important in aerating the soil and in assisting in the recycling of plant materials. The insects, a group which evolved at about the same time as the flowering plants exploit a plant diet more than other invertebrates. Insect herbivores range from small soil living springtails living on

40

decaying plant material and pollen grains to locusts eating their own body weight of plant material each day and to plant parasites such as aphids and bugs. Butterflies and moths have different diets in their larval and adult stages. Caterpillars may be specific to one or two food plants, often because the parent butterfly chooses to lay her eggs on those plants, and voraciously eat the food plant on which they develop. Adult butterflies sip nectar. Bees also feed on nectar and collect it in a honey sac in front of the stomach. In the hive this is disgorged into a special cell, secretions are added and the mixture thickens by evaporation of water to become honey. An active hive can produce about 2 lb (0·9 kg) per day in summer, this representing some 5 million bee journeys to plants. A dusting of pollen is collected on the hairy bodies of the bees and is groomed off into special pollen baskets on the hind legs; this provides another, protein rich source of food. The Ambrosia beetles and Leaf-cutter ants of the New World cultivate fungi which form their main diet. The Leaf-cutters bite off large pieces of leaves and petals which are chewed but not eaten.

These are deposited as a kind of compost on the floor of part of the nest and fungi grow on this, perhaps aided by growth promoting secretions from the ant's saliva present in the chewed leaves. The ants devour the bulb-like outgrowths produced by the fungus when tended by ants but not otherwise. When moving nests the ants may carry a piece of fungus with them to start the new fungus garden.

Many shore dwelling animals live on sea weeds and algal films. Sea slugs, like the Sea hare already mentioned, may also match the weed they eat and some actually store chloroplasts in their own tissues. Limpets, chitons or Coat of mail shells and Sea urchins all live by scraping off and eating the algal slime from rocks. They all have strong rasping teeth to do this. In the limpets and chitons is a remarkable long tongue called a radula covered with teeth which is continuously produced at one end as the other is worn away. Teeth on the radula of chitons are strengthened with iron salts. Sea urchins have a circle of five hard teeth each made of four articulating pieces and operated by powerful

Geometric shapes of phytoplankton from the English channel which contains various types of diatoms and dinoflagellates. These microscopic plant cells are the basis of all life in the sea.

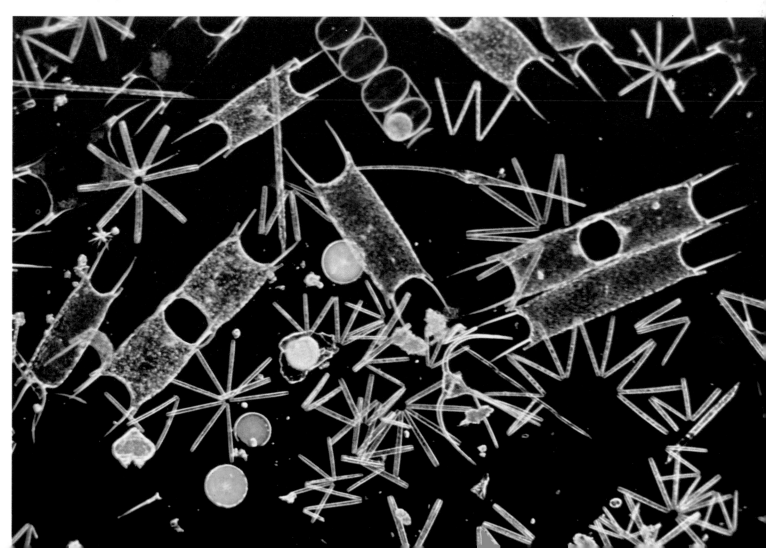

This tiny marine shrimp *Calanus helgolandicus* is an extremely important member of the zooplankton because it is the main food of herrings and Basking sharks. *Calanus* itself is a filter feeder and uses the vibrations of its long antennae and mouthparts to collect food.

muscles; this apparatus is called Aristotle's Lantern. Tube feet around the mouth help to push food into the mouth and are also sensory.

Particle Feeders. The illuminated surface waters of lakes and seas contain a rich soup of single celled diatoms and dinoflagellates. This is fed upon by the primary consumers, many of which are the small larvae of sedentary invertebrates that are set adrift for a while to exploit these rich pastures. Adult invertebrates, particularly crustaceans, are also specialized plankton feeders. A small copepod shrimp, *Calanus*, which is only about 0·2 in (5 mm) long but occurs in enormous numbers, is an extremely important protein producer in marine food webs and is eaten extensively by fishes such as herrings and by basking sharks. *Calanus* is a filter feeder and uses vibrations of its antennae and mouthparts to create water vortices bringing food. The water currents are directed through hair-like filters, the stationary maxillary appendages, which strain out food and about 8·2 cu in (135 ml) of sea water can be sieved each day from which organic matter is extracted. The shrimp-like euphausids known as krill which are larger than *Calanus* and measure 1–2 in (2–5 cm) long are also important primary consumers and feed basically on phytoplankton but may also eat copepods. Translucent, barrel-shaped salps which are related to Sea squirts are also pelagic filter feeders and have a series of fine pharyngeal ports and screens through which minute

protozoans and algae are filtered out. Zoologists have exploited this efficiency by collecting rare protozoans and diatoms from salp stomachs for study: these being too small to be collected using even the finest meshed silk plankton nets!

Apart from these drifting plankton feeders, particle feeding is also very much associated with a sedentary way of life since animals attached to rocks or buried in sand cannot hunt for food but have to create currents to bring food to them. The barnacle inside its volcano of shell plates has legs covered with fine bristles which act as a kind of grab net to collect small animals and plants. Tube worms, bivalves, Sea lilies and Sea squirts have special tracts of lash-like cilia which create feeding currents and also act as definite conveyor belts for food particles which are consolidated with sticky mucus secreted from the food collecting surface. The fanworms are amongst the most beautiful of the tube worms. They live in long tubes that project from the surface of the mud and have a crown of feathery colour-banded processes which bear cilia that create feeding currents. As this crown is a tempting morsel for fishes it can be withdrawn very rapidly into the tube when a shadow falls on eyespots associated with it. The crown can also be regenerated rapidly if bitten off or damaged. The food particles are sorted according to size at the base of the crown; the smallest, consisting largely of algae, are passed to the mouth and eaten and the

These mediterranean serpulid tube worms collect fine particles of food from the water by means of the brilliant red crown of tentacles which set up water currents. The worms can withdraw into their shells, which are encrusted tubes, for protection if threatened.

The fanworms have a feathery crown of tentacles on their heads which collect and sieve out small living organisms from the water. This fanworm from the coast of northern Spain has a spiral crown.

largest are rejected by means of ciliary paths leading away from the mouth or may be mixed with mucus and added to the smooth tube composed of mud. Bivalve molluscs such as mussels, scallops and oysters are highly developed filter feeders and use the ciliated surfaces of highly expanded gills to collect food particles. Burrowing bivalves such as clams have long siphons stretching to the top of the mud through which water containing food is drawn. Ciliary tracts propel particles down the gills from the hinge region to the free edge, from here along the free edge and then to the large flap-like labial palps on each side of the mouth where a final sorting occurs. The particles are consolidated by sticky mucus secreted by the gills and become bound into strings. The ciliary sorting mechanism doesn't stop at the labial palps but continues in the stomach which contains ciliary tracts. Also present in the stomach is a so-called crystalline style, a translucent

rod of mucoprotein containing an enzyme that digests starchy substances. This is rotated slowly by cilia and mucus food strings are wound on to it and hauled into the stomach as though on a windlass. The tip of the crystalline style rubs on a hard shield which abrades it and releases the enzyme. This digests the food particles whilst the acidity of the stomach releases food particles from the mucus string. Only very fine food particles are allowed into the digestive glands and it is here that the ciliary sorting paths in the stomach come in. Most bivalves feed on plankton, bacteria and detritus and different kinds of oyster can filter between 2·2–7·5 gal 10–34 l) of water per hour which can produce 0·35 oz (1 gm) of food per day. There is some evidence that the rate at which oysters pump water is influenced by the concentration of carbohydrates released by the phytoplankton on which the oyster feeds. It is difficult to decide whether bivalves filter continuously because although water may pass into the shell most of the time, mucus is not always secreted on the gills, in which case feeding is probably not occurring. The oyster filter feeds more or less continuously throughout life and has a dispersive transparent larva which uses cilia on its two head lobes both for swimming and to collect food particles. Many of the fragile and diaphanous drifting larvae of the zooplankton feed in this way. Some commensal animals take advantage of the efficiency of adult shellfish in collecting food and take up residence in their shells. Small Pea crabs are extremely common and have special hair fringed claws to pick food-containing mucous strings off the gills of the host. A handful of marine animals have perfected ways of using sticky mucus alone as a kind of fly paper to catch small organisms. The marine snail *Vermetus*, which has a shell shaped like a partially unwound spring, does this. It secretes a 12 in (30 cm) long train of mucus from a gland in the foot; this traps food and is hauled in every so often and eaten.

Predators. Predatory animals have to have sense organs to detect their prey whether vibration receptors (as in marine arrow worms and spiders), eyes or olfactory organs. They may have to pursue, stalk or ambush their prey and are often camouflaged to avoid arousing the suspicions of the prey until capture has been made. A predator needs probing and grasping organs such as clawed legs, proboscides and tentacles, jaws for biting and holding and venoms and anaesthetics to kill or immobilize the victim. As the predatory mode of life is more hazardous than that of a herbivore or a filter feeder the ability to fast for long periods, or to make extensive use of their own tissues is an advantage. The Sea anemones, corals and jellyfishes (cnidarians) have a unique adaptation for catching prey, namely their stinging cells or nematocysts. These are present mainly in the tentacles but are also found on filaments present in the digestive cavity and in skin covering the body. Each stinging cell consists of an elaborately coiled, often barbed tube wound within a double walled capsule. When the nematocyst is stimulated by touch or by chemicals from the prey the thread is shot out through the lid at the top of the capsule, turning inside out as it does so. The thread rotates counter clockwise like a drill as it is fired to penetrate the prey and inject venom. Release of the thread may be brought about either by direct tactile or chemical stimulation of a sensory process on the nematocyst itself or by stimulation of a neighbouring sense cell. The mechanism of ejection is not known but it may involve a hydrostatic driving force. Each nematocyst can be discharged only once so new nematocysts have to

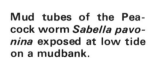

Mud tubes of the Peacock worm *Sabella pavonina* exposed at low tide on a mudbank.

The cockle cannot chase its prey and so lies half buried in the sand and sucks water into its shell by means of a pipe-like siphon. The water is filtered through the cockle's gills inside the shell and suspended food matter is removed at the same time as oxygen is taken out of the water. The water is then pumped out via another siphon.

be produced continuously. In addition to about six different kinds of stinging cell, Sea anemones and Stony corals have cells containing sticky cells termed spirocysts. The nature of the stinging cell venom employed by cnidarians is not yet fully worked out but some may be simple proteins. Many anemones and corals wait for small animals to brush against their outstretched tentacles, but some jelly-fish show fishing behaviour and swim to the surface and then float downwards with the tentacles outspread to catch prey. Forms like the Portuguese man-o'-war look like jellyfish but are actually a whole colony of individual polyps hanging beneath a gas filled float. The stinging tentacles belong to one kind of polyp and when a fish swims against them the tentacles contract bringing the fish in

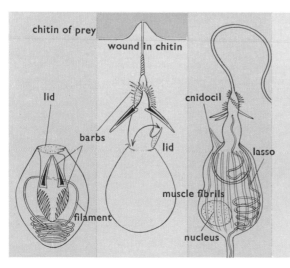

The stinging cells of Sea anemones, corals, jellyfish and *Hydra* consist of a coiled thread bearing barbs which can be shot out of a capsule when a trigger is stimulated. There are several kinds of stinging cell in the same animal and some inject paralysants and venoms into their prey. A sting cell is shown undischarged (left), discharged (centre), and partially discharged with some thread still within the capsule (right).

Marine predators. The innocuous looking Beadlet anemone has tentacles packed with stinging cells which catch and paralyse small animals in the water. The white and yellow rough shelled Dog whelks feed on mussels by rasping a hole in the shell and feeding on the soft contents.

contact with another kind of polyp immediately under the float that digests the prey and passes nutrients into the communal body cavity.

The stinging cells of cnidarians do not make them immune to predators. Certain Sea slugs find various Sea anemones extremely palatable. The European grey Sea slug *Aeolidia papillosa* feeds on the Beadlet anemone and the Plumose anemone which are much larger than itself. Interestingly the aeolid Sea slugs, themselves unaffected by Sea anemone stinging cells, disguise themselves as anemones. They have long tentacular outgrowths or cerata on their backs and actually incorporate undischarged stinging cells obtained from their prey into these processes. The stinging cells are sorted in the gut of the Sea slug and somehow transported without being discharged into the cerata. In some cases only the most dangerous kind of nematocyst is selected for incorporation into the cerata and interestingly the anemone does not discharge these when being fed upon by the Sea slug. Another predator of cnidarians is the Crown of Thorns starfish that grazes on corals and devastates coral reefs, leaving bare white skeletons.

Some of the Sea snails to which the Sea slugs are distantly related are amongst the most voracious of marine carnivores. Particularly interesting are those that feed on their fellow Sea snails or on bivalves by means of a long proboscis with file-like teeth at the end. The Dog whelk uses its proboscis to rasp a round hole in the shell of a mussel or inserts it into the shell plates of a barnacle, so that the soft parts can be scraped out. Some whelks cleverly use dor-to-door salesman techniques and jam their own shells into the opening of, for instance, oysters and then insert their proboscis into the soft parts. *Natica*, the Necklace shell, produces sulphuric acid from the proboscis glands to assist in boring through the shells of its prey. One might imagine that an oyster inside its thick shell would be fairly safe from predators, but in fact it is attacked by a whelk called the Oyster drill *Urosalpinx* a bristle-worm *Polydora* that bores into the shell, a boring sponge, a flatworm *Pseudostylacus* and starfish that pull the valves apart using their arms. Oysters are also sensitive to competition from other sedentary filter feeding animals competing for space and food.

Crepidula, the Slipper limpet, and barnacles are all rivals in this respect.

It is difficult to realize that the cuttlefish, squids and octopuses are molluscs and belong to the same group as the Sea snails and oysters. These highly mobile and intelligent animals use their sucker-bearing arms and tentacles to catch prey. Squids feed mainly on fish and can make a dash at speed using jet propulsion. Two long tentacles each bearing a pad with suckers at the end are shot out from amongst the eight arms to trap fish. Cuttlefish catch food from a position half buried in sand during the day. The cuttle camouflages itself to match this background using pigment bags called chromatophores in the skin. These are under nervous control and can expand and contract very rapidly sending waves of colour shimmering over the body surface. From this concealed position the cuttle beckons with the two uppermost arms as a lure for curious prawns and crabs. As these approach they are captured by the two long tentacles that shoot out from amongst the arms. In the excitement of capture the cuttle may change colour completely – it seems to be a very emotional animal. Octopuses catch prey on the sea bed and live in cairns of stones or in caves. Prey is seized by the curling arms as it approaches. Many octopuses are inordinately fond of crabs and these are used as 'bait' in conditioning experiments on octopuses. All cephalopods, the groups to which squids, cuttlefish and octopuses belong, have a horny beak for biting as well as a file-like tongue, the radula, for tearing flesh. The salivary glands contain venom consisting of a nerve poison and enzymes which digest proteins and these are injected during biting.

◁ Cuttlefish showing the striped colour pattern used during courting. The long tentacles used for snatching prey are hidden within the 8 arms. The cuttle has well developed eyes for sighting prey.

The round siphonophore *Porpita* ▷ looks like a jellyfish but is really a colony of polyps hanging below a disc-shaped gas-filled float. The stinging cells of this animal are no protection against the floating sea slug *Glaucus* which is seen eating it. The sea slug is bottom left and has feathery processes on its body to support it in the water.

Land Predators – Nets and Traps. There are many notable predators on land ranging from the *Scolopendra* centipedes of the tropical Americas, which may grow up to 1 ft (0·3 m) long, have a great fringe of legs around the body and feed on mice using poison claws to immobilize their prey, to forms like the Army ants. These are perpetually on the move and, although they usually feed on small insects, will devour tethered horses and glutted snakes that lie in the path of a migrating column. Perhaps the most interesting of the land and freshwater predators, however, are those that weave silken traps to ensnare their prey. This elegant method of predation is not confined to spiders but also occurs in certain insects. One fresh water Caddis fly *Hydropsyche* weaves a seine net which is strung across fast flowing water and anchored to stones on each side. At intervals the net is inspected for aquatic insects and other prey that the current has swept into it. Some cave and crevice living Fungus gnats of New Zealand have larvae that catch other insects using luminous lures. The gnat larvae are about 1 in (3 cm) long and secrete a horizontal supporting thread from which are suspended a number of trap threads 1–15 in (3–50 cm) long covered with sticky droplets. The larva lies along the horizontal supporting thread and illuminates the hanging trap threads using its luminous tail end. When an insect flies into the sticky threads it is reeled in and eaten and the trap is then repaired. The Fungus gnat larvae occur in large numbers in some caves and produce an ethereal blue star spangled effect on the cave roof. For sheer artistry in capture technique, however, the spiders can hardly be bettered. The Orb web spinners which make a wheel-shaped web supported by radial spokes are commonly encountered in gardens. The female spider sits head down at the centre of her web, her eight legs spread to detect the direction of vibration caused by an entangled insect. The centre or hub of the web on which the spider sits is formed of meshed threads surrounded by a strengthening spiral of six or seven turns. This centre region is not sticky but outside this the twenty to forty turns of spiral thread are covered with small droplets of glue which hold the insects that blunder into the web. The spider is prevented from sticking to her own web by means of a thin film of oil on her legs. If this is dissolved off

The Garden spider sits head-down in the centre of her orb web, her tight legs spread to detect the direction of vibration caused by struggling insects. The central part of the web is not sticky but the outer turns of thread are covered with droplets of sticky glue.

A Sheet web spider waits for insects to fall into the closely woven hammock-like net after they have tangled with the support thread above.

using chloroform the spider can be caught in her own web. The spider also has the tip of the third leg modified into a grappling hook through which run the silken strands of the web so that the spider can run quickly yet securely over the surface of the web. Once the vibrations of a trapped insect are felt, the spider runs towards it, bites it and enshrouds it in silk. Digestive enzymes are injected into the prey via the fangs and dissolve the tissues which are later sucked out by the spider. Not all prey is acceptable to the spider. Cinnabar and Burnet moths apparently taste nasty and are often rejected and cut out of the web after being bitten. They may survive this and probably have some immunity to spider

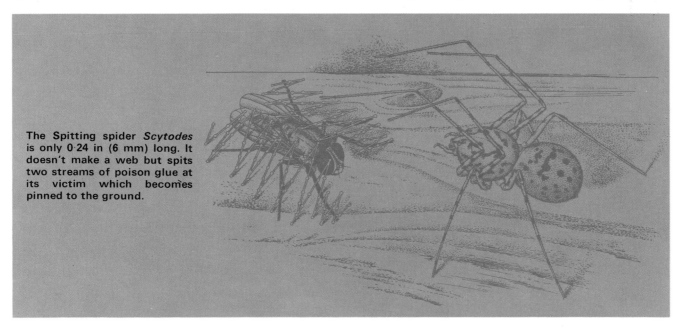

The Spitting spider *Scytodes* is only 0·24 in (6 mm) long. It doesn't make a web but spits two streams of poison glue at its victim which becomes pinned to the ground.

The Crab spider *Miscumeria* does not make a web but simply sits in a flower and waits for an insect visitor. The crab spider is usually white, pale pink or yellow to match its floral trap and so remains invisible to its prey until too late.

venom and enzymes. Some pompilid Spider hunting wasps apparently produce a warning scent which may be recognized innately by many spiders. When a Garden spider is attracted to and touches a pompilid trapped in its web it soon beats a hasty retreat. One aberrant tropical Orb spider has

The European water spider spins a silken underwater diving bell for storing air. From this base it preys on underwater insects and crustaceans.

The whelk *Buccinum undatum* attacking a Carpet shell *Venerupis pullastra*.

abandoned making a web and has taken to whirling a single thread with a blob of sticky glue at the end like a bolas, to trap insects. Other kinds of silken traps woven by spiders are the sheet-like hammock webs which are very conspicuous when dew laden in autumn filming large areas of gorse and grass. These are woven by small black shiny spiders often called Money spiders. Sheet webs catch hopping and other insects that tangle with the support threads above the net and fall into the hammock spread below. The owner of the net lurks under this and bites the prey through the net before tugging it through and wrapping it in silk. Purse web spiders have a similar method of attack. They produce a cylindrical web which has an underground tunnel portion and an outer purse-shaped end which lies along the ground. The spider sits on the roof of the outer web and waits for the vibrations that signal a capture. Running to the point below the hapless insect the spider thrusts its long fangs through the web into the prey and injects its venom. A slit is then made in the web, the prey dragged through and the net then repaired. Sprung traps are another kind of specialist web. The British heathland spider *Achaearanea* uses a sprung trap to feed on ants. From a tangled superstructure a series of tensioned threads run vertically to the ground, each bearing a sticky droplet at the base. This trap is designed to catch ants and as an ant blunders into the web it breaks the thread and is jerked into the air. The spider then descends to deal with its struggling prey. Not all spiders use webs as traps. The remarkable Spitting spider *Scytodes* actually spits two streams of poison glue at its victim which becomes pinned to the ground. Hairy Bird-eating spiders which are as large as a man's hand hide in crevices during the day and at night dash after small mammals or humming birds. They discourage their own predators by being covered with irritant hairs that give them their furry appearance. The nomadic Wolf spiders freelance without webs, pursuing their prey at speed even when carrying young spiders that cling to the back of the female. Jumping spiders stalk their prey and have very accurate vision to gauge the distance of the leap they must make onto their victim. Crab spiders use natural traps to catch insects and simply sit motionless inside a flower waiting for the visit of an insect pollinator. The irregular shape of the Crab spider, together with its often bright colouration, helps it to blend in with the

53

Black aphids suck plant juices and are in turn milked for their honey dew by Wood ants.

background. A British Crab spider that lives inside the pink bells of heather is itself a remarkable pink colour; the female Crab spider *Misumena vatia* can actually change its colour to match the white or yellow flowers in which it lives and is said to do this by pumping a yellow liquid pigment from the intestine to the skin. A white colouration is resumed on withdrawal of the pigment. Many Crab spiders specialize on a diet of bees; these being amongst the commonest of the pollinators. Spiders, like most successful animal groups, have their specialists. Different species prey in woodland, in fields and on the sea shore and different types have evolved various ways of catching specific kinds of insects. Notable amongst these specialists is the Water spider which has taken over a whole new environment. By filling a silken net with air bubbles, the water spider creates a silvery diving bell which enables it to live in ponds and lakes. The thimble-shaped diving bell, which is anchored to weeds, forms the base for the spider which ventures out to catch aquatic insects and crustaceans that come within range. At night the spider moves further afield but always has to bring prey back to the diving bell to be injected with digestive juices and later sucked dry.

Animals in Association – Parasites, Symbionts and Commensals

Almost every animal, whether a protozoan, sponge, Sea anemone, mollusc or mammal, has other organisms of a different kind that live in association with it. The partners involved may be very dissimilar, so that bacteria, algae and protozoans may live in the gut or tissues of a larger multicellular animal or two or more multicellular animals may live in association. These relationships between species usually involve particular species of partners and can show a high degree of intimacy and specialization so that one of the partners at least may have become completely dependent on its associate for its livelihood. The structural and behavioural adaptations shown by one or more of the partners to the shared way of life indicate that many of these associations are very old and well established and that the two organisms may be evolving as a unit. Relationships of this kind form a special aspect of ecology since the larger partner, or its immediate surroundings, form at least temporarily the environ-

ment for the smaller partner or partners. Similarly the smaller members of the partnership become an environmental feature of the larger organism. The relationship may involve several partners. The burrow of an echuroid worm living off the coast of North America is shared with a goby, a Pea crab, a Scale worm and a small clam. It is hardly surprising this worm is sometimes called the 'inn keeper' because of all the guests it harbours in its burrow. To adapt a comment of A. E. Shipley's, one might say that some animals are veritable perambulating zoological (or botanical) gardens because a single animal may have an amazing number of organisms living in or on it. Those organisms which are entirely dependent on their hosts are called parasites. The so-called best friend of man, the dog, can carry some 300 parasites and about 50 of these, some extremely debilitating, can be transmitted to man. Each particular relationship has its own raison d'être and own characteristics and as these

Red mites hitch a lift from a harvestman by clinging to its body. This kind of transport association is known as phoresy.

The Dor beetle *Geotrupes stercorarius* with mites *Gamasus coleopterorum* riding on it. In evolutionary terms it is often a short step from a harmless association like this to a truly parasitic relationship.

associations are very common in nature, each kind of relationship shades into other kinds so that there is a continuous spectrum of give and take. In addition the nature of the relationship is constantly changing both throughout the life histories of the partners and more gradually due to evolution. This means that it is very difficult to classify or define any particular relationship. Some animals are, for instance, parasitic only during their larval stage (protelan parasites). A few animals have some generations that are parasitic and some that are free living. For instance, there is a nematode worm called *Rhabdias bufonis* that can either live in the lungs of frogs and toads, where it reproduces asexually by a process called parthenogenesis, or can have generations that are free living in soil and reproduce sexually. The physiological switch mechanism that determines whether the worm is to be free living or parasitic is not yet understood. Animals that do not depend entirely on parasitism are called facultative parasites. A particularly good definition-defying example of the kind of complexity that can exist in relationships between animals is provided by the partnership of a small carp-like fish, the bitterling *Rhodeus sericeus* with freshwater mussels *Unio* and a species of *Anodonta*. As will be mentioned later, the bitterling is completely dependent on the mussel for protection of its eggs and fry. It spawns into the mussel and will not spawn unless a mussel is present. The mussel is not a passive partner in this relationship and uses the bitterlings as hosts for its parasitic larvae. This

relationship broadly benefits both partners but involves one partner (the bitterling) being completely dependent on the other (the mussel) which actually parasitises it!

Broadly speaking the following kinds of relationship can be recognized. In relationships where both partners benefit, the association is called symbiosis (from the greek syn, together; bios, livelihood) or mutualism. There is some confusion over the precise usage of the words symbiosis and mutualism but the two terms will be used interchangeably here. Where one partner benefits at the expense of the other and may actually do the other (the host) some harm, the relationship is termed parasitism (greek para, beside; sitos, food). If there is a regular and close association between two organisms where only one partner benefits but the other does not suffer any serious disadvantage, the relationship is termed commensalism (latin cum, with; mensa, table). Commensalism often involves a smaller species sharing food with a larger one and being a kind of messmate, or living in association with it for protection. Another kind of relationship which can be just as specific as any of those so far mentioned involves one species using another as a means of transport but without feeding upon its partner's tissues. This is called phoresy (greek, pherein, to bear).

Phoresy – Animal Hitch-hikers. The freshwater shrimp *Gammarus* has three kinds of hitch-hikers on its limbs. There are at least two kinds of protozoan, a large suctorian *Dendrocometes paradoxus*

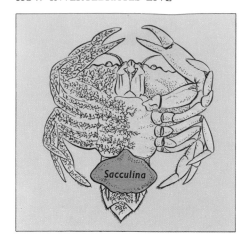

The crustacean parasite of crabs, *Sacculina*, has lost its nervous system and limbs and turned into little more than a branching rootlet system within its host tissues. The dark brown area is the external reproductive sac of the parasite.

and a species of flagellate *Conotricha* and rotifers *Proales parasitica*. These organisms probably use *Gammarus* not only for transport but also to provide feeding currents. They must be very dependent upon the shrimp because they do not occur on their own. Other kinds of aquatic crustaceans have similar protozoans associated with them. The barnacles that attach to whales provide another example of phoresy. One of the best known, *Coronula*, is related to the common Acorn barnacles that attach to rocks but is much larger. Its larva may be specifically attracted to whale skin since *Coronula* occurs only on them and is deeply bedded into the skin. Another kind of Whale barnacle, *Conchoderma auritum*, this time related to the Goose barnacles with a long stalk, is often found fastened to *Coronula* although it can also attach directly to whale skin.

There are in fact a whole series of different kinds of barnacles that live on other animals and whilst a few are phoretic some have become parasites such as *Rhizolepas*, a highly modified barnacle that sends long branching rootlets into the tissues of its annelid host. The rootlets are used to absorb nutrients. Another highly specialized parasite related to the barnacles is a species of *Sacculina* that lives on crabs, which is so highly modified that it can be identified as a barnacle only by its horned nauplius larvae. The adult parasite consists merely of a branching rootlet system which ramifies throughout the tissues of the crab and an external reproductive sac. Another example of phoresy that shades into parasitism is provided by the tiny Sea slug *Phyllirhoë* that hitches a lift from a small jellyfish *Zanclea*. As the Sea slug grows, it starts to eat the jellyfish stage, and when it is eventually independent and dwarfs the latter completely it finishes off the whole jellyfish, tentacles and all. The term phoresy was once used specifically to designate the transport of small arthropods by insects and it is well known that beetles and other familiar insects are often found to be covered with mites.

Commensalism – Messmates and Protective Associations. There are many examples of smaller animals taking shelter with a larger animal, particularly inside its shell, tube or burrow and at the same time benefiting from scraps of food left over from the host's meal and from oxygen carrying water currents created by the host to irrigate its own surroundings. Some of the best known examples of

◁ The large sac-like structure under the abdomen of this crab is the reproductive part of a parasite, *Sacculina*. The rest of the parasite is not visible for it forms an absorptive branching network inside the crab's tissues. Distantly related to the barnacles this parasite has become extremely modified in adaption to its way of life.

Clownfish *Amphiprion percula* ▷ lives in symbiosis with Sea anemones protected from predators by the stinging tentacles of its partner. The Sea anemone may benefit from scraps of food that the fish either brings to it or accidentally drops.

this are provided by marine bristleworms, species of which associate specifically with Hermit crabs, prawns, clams, Sea stars, Sea urchins and Sea cucumbers. The relationship between the large ragworm *Nereis fucata* and the Hermit crab *Eupagurus bernhardus* has been well studied. The worm lives in the upper whorls of the whelk shell occupied by the Hermit and maintains a water current for respiration by pulsations of its body. When the Hermit is feeding the worm creeps out and snatches pieces of food from its mouthparts, retracting quickly with its stolen prize. The behaviour of the worm has been studied by making both Hermit and worm accept an artificial glass shell through which the movements of worms can be seen easily. Young worms are found in muddy tubes on the sea bed and they are able to contact a suitable host by being sensitive to vibrations similar to those produced by a Hermit bumping its shell along.

The freshwater oligochaete worm *Chaetogaster* has evolved a similar kind of habit to that of *Nereis fucata*, but lives in the occupied shells of freshwater snails. There are reports that this commensal eats fluke larvae that are either about to colonize or are being produced and liberated by the snail and it has been suggested that *Chaetogaster* might be useful in the biological control of fluke diseases.

Several Scale worms live commensally in the burrows of other marine animals. One called *Hesperonoë* lives with an American prawn *Calianassa californiensis* and eats scraps drawn into the burrow. It has fierce territorial instincts and drives off other Scale worms that intend to intrude upon the twosome. Some rather daring Scale worms also live on the Sea star *Astropecten irregularis* and have been observed to insert up to a third of their body into the host's stomach searching for food. The Scale worms seem to respond to a specific chemical attractant produced by Sea stars which enables them to recognize suitable partners. They are said to be repelled by wounded or dying hosts and presumably seek out another partner in such cases.

Animals that filter feed often prove attractive hosts for commensal associates because the food collection region is often situated in a protected vestibule inside the animal which is convenient for plunder. For this reason sponges, Sea squirts and bivalves often harbour many inmates. Various kinds of shrimps and worms occur inside some sponges and in one well known example the relationship becomes more intimate than the shrimps might have wished. This concerns the beautiful siliceous Venus flower basket sponge which shelters pairs of shrimps *Spongicola* in its water passages. As the shrimps grow they become too large to escape and are doomed to remain until death in their spun 'glass' prison. The Japanese collect the skeletons of the sponges which occur off their coasts and give them as wedding presents where the entombed shrimps inside symbolize the permanence of the marriage bond. Even small invertebrates may

Commensal relationship between the Hermit crab *Eupagurus bernhardus* and the ragworm *Nereis fucata* clearly revealed by giving the crab an artifical glass shell. The worm lives in the upper whorls of shell well protected by the crab and occasionally emerges to snatch food.

Symbiotic association between three Sea anemones *Caluactis parasitica* and a Hermit crab *Eupagurus bernhardus* living in an empty whelk shell. The anemones protect the crab with their stinging tentacles whilst the anemones probably benefit from scraps of food left over from the crab's meal.

harbour an amazing number of cohabitants. A species of Sea squirt *Ascidia* 2·75 in (7 cm) long was found to contain two different kinds of bivalve, a Pea crab, an amphipod shrimp and two species of copepod shrimps. The Pea crab that lives inside bivalves and picks up food strings from the host's gill margins with its claws, has already been mentioned. *Pinnotheres pisum*, the small crab that associates with mussels, does them little harm but the Pea crab, *Pinnotheres ostreum*, that lives in the American oyster *Crassostrea* damages the oyster's gills and so can be considered as a parasite.

The art of Protection and Disguise. The stinging animals, Sea anemones, jellyfish and corals, enter into interesting relationships with other animals which are protected by them. The example of the Clown fish that live amongst the tentacles of giant Sea anemones on coral reefs is well known. Often a pair of Clown fish will adopt a single anemone which then becomes their territory. Some fish are specific to one kind of anemone but others can colonize several species. The fish swim amongst the tentacles of the anemone without being harmed although other fish of the same size will be stung and eaten. The immunity of the Clown fish is built up gradually. Firstly the fish chooses the right kind of anemone. It then 'acclimates' the anemone by swimming slowly close to the anemone in a par-

ticular manner. When the fish first contacts the anemone's tentacles, these stick to the fish which has to jerk violently to free itself. As the fish repeatedly contacts the tentacles these cling less and less to the fish until eventually the fish evokes no response from the anemone and is free to swim safely amongst the tentacles. The recognition or 'acclimation' process takes about an hour and experiments in aquaria have shown that if the 'acclimation' ritual is not followed, Clown fish may be captured and eaten. The fact that the anemone can learn to recognize the Clown fish shows that completely different kinds of animals can learn to communicate with one another using chemical, tactile and behavioural 'language'. This ability to communicate forms an essential part of intimate relationships of this kind. A substance has been isolated from the mucus on the fish's skin that raises the threshold of mechanical stimulation of the anemone's stinging cells so that these cannot be discharged as easily. The association obviously benefits the fish because if they are separated from their anemones and placed in aquaria with other fish they are often eaten. The advantage gained by the anemone is less clear; however, the fish may act as decoys and lure prey within reach of the anemone. They may scavenge the waste food material of the anemone or even bring food to the anemone and

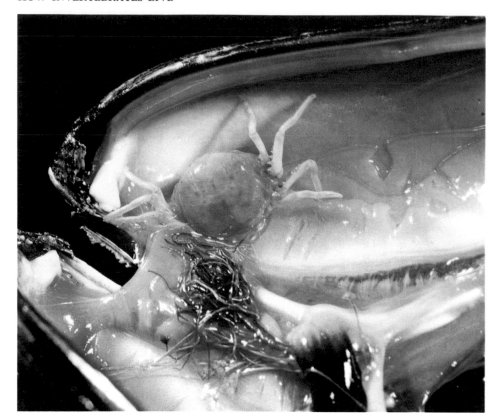

A Common mussel opened to show a pink bodied Pea crab living commensally in the mantle cavity. The Pea crab does not harm the mussel but uses its hairy claws to thieve food from the mussel's gills.

drop it amongst the tentacles. This has been reported for the Clown fish *Amphiprion polymnus*, but whether it happens regularly in nature is not known. This particular relationship may therefore be symbiotic rather than commensal.

A surprising number of fish also gain protection by sheltering under the umbrellas of jellyfish amongst the stinging tentacles. Fish fry which are extremely vulnerable are particularly prone to this and in some cases the relationship is obligatory and the fish will die unless a suitable jellyfish is found. The fry of whiting, cod, haddock and Horse mackerel have all been recorded to form associations of this kind. A small blue and silver striped fish, *Nomeus gronovii*, is associated for much of its life with the Portuguese man-o'-war and darts precariously amongst its stinging tentacles. The fish has some immunity to the toxins of its host and is adequately protected from predators by the deadly strands that hang around it. Various kinds of crabs show highly specific and presumably long established protective relationships with Sea anemones and encrusting hydrozoans. The anemone *Calliactis parasitica* is frequently found attached to the whelk shells adopted by Hermit crabs *Eupagurus bernhardus*, *E. striatus* and *E. arossor*. Sometimes several anemones may crowd onto one shell. The

anemone can live alone and is also sometimes found on empty whelk shells but in experimental situations will tend to transfer to a shell occupied by a Hermit if one is available. An anemone already living with a Hermit will change shells with the crab when the Hermit outgrows one and moves to a larger one and in some cases the transfer is assisted by the crabs. It has been shown that the anemone recognizes chemicals in the periostracum or organic coat of the whelk shell and that this provides a powerful stimulus for attachment. The anemone first contacts the shell with its tentacles and having recognized the attractant, nematocysts are discharged that attach the tentacles to the shell. The anemone then brings its foot into contact with the shell and hauls itself into position. In the case of *E. bernhardus* the crab makes no effort to assist the anemone but the other crabs mentioned, *E. striatus* and *E. arossor*, will seize *Calliactis* with their claws and manhandle it onto the shell if they find themselves without an anemone partner. This determined behaviour on the part of the Hermits does suggest that there is some positive advantange to be gained from the presence of the anemones. Almost certainly the crab is protected by the stinging cells of the anemone and may even be camouflaged too. The Mediterranean Hermit crab *Dardanus* has in fact recently been shown to be protected from the attacks of *Octopus* by the anemones *Calliactis* on its adopted shell. Although this relationship is often referred to as commensal it is probably actually symbiotic because the anemone is thought to obtain food from the crab partner and bends down to sweep the ground with its tentacles when the crab is feeding. Another famous Hermit crab-anemone team is the association of *Eupagurus prideauxi* with the anemone *Adamsia palliata*. The Hermit crab in this association always chooses to live in a shell that is too small for it and the anemone comes in to fill the gap. It wraps its foot around the body of the Hermit, completely covering the adopted shell. From the back of the crab the tentacles of the anemone can hardly be seen because these are

positioned under the crab's body and just behind the mouth in an ideal position for food sharing. It appears that the relationship, which is obviously of benefit to both partners and constitutes a symbioses are part of a wide spectrum of crab-occasionally places food amongst the tentacles of the anemone. The relationship is highly specific; *Adamsia palliata* is never found without the Hermit crab and indeed once both are past their juvenile stage neither can exist without the other. If the crab is removed from its shell the anemone drops off and soon dies. These particular examples of protective symbioses are part of a wide spectrum of crab anemone relationships. Some remarkable coral reef crabs, for instance *Melia*, grab small anemones from rocks when they encounter them and carry

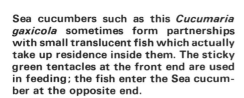

Sea cucumbers such as this *Cucumaria gaxicola* sometimes form partnerships with small translucent fish which actually take up residence inside them. The sticky green tentacles at the front end are used in feeding; the fish enter the Sea cucumber at the opposite end.

them around in their claws to brandish at enemies. They also appear to use the anemones as a food collection device and will rob the anemones of any food they catch by picking it out of their tentacles. The armoury of another kind of invertebrate, the Sea urchin *Diadema* is exploited for protection by two completely different species of fish. This black sea urchin has long fragile spines with poisonous

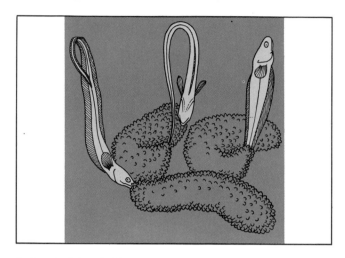

A Pearl fish of the genus *Carapus* shown entering and leaving a Sea cucumber which it uses as its home. The Pearl fish goes into the Sea cucumber tail first but comes out head first in case of danger.

tips that easily break off and injure an attacker. Nevertheless it has become home for the tropical Shrimp fish and Cling fish which hover head down between the spines. Remarkably both have evolved an elongated body with camouflaging stripe of contrasting colour running the full length of the body so that they are quite difficult to see through the forest of spines and are anyway almost impossible to attack. Sea cucumbers lack spines but some species contain powerful poisons in the leathery body wall and others discharge sticky noxious mucous threads that discourage attackers. These effective defence mechanisms have led to Sea cucumbers being adopted as citadels by certain small translucent fish called pearlfish *Carapus*. The fish enters the Sea cucumber tail first via the anus and shelters inside. The Sea cucumber breathes through the anus so this is never closed for long. The relationship is obligatory because the young fish need to find a Sea cucumber before they can mature. The fish have become especially modified for this strange partnership by losing their scales and pelvic fins which allows them to offer little resistance as they slip backwards into the sea

cucumber. The anus of the fish has also shifted far forward so that it can void its faecal matter outside the body of the Sea cucumber. When the fish wants to enter the anus of the Sea cucumber it 'taps at the door' by nudging the back end of the Sea cucumber with its head. As soon as the host opens up the fish whips round and inserts its long slender tail. Whilst some species of pearlfish appear to do their hosts little harm, others are exploiting the relationship in the direction of parasitism. The Mediterranean pearlfish *Carapus acus*, for instance, breaks through the cloacal wall into the body cavity of its Sea cucumber partner and browses on the gonads. This again shows how delicately poised these kinds of relationships are and how they can easily change in character.

Cleaning Symbioses. Many different kinds of animals have independently evolved the habit of cleaning or grooming animals of a different species and removing and eating the parasites they find on them. Birds such as the Cattle egrets have taken up residence with cattle, antelopes and rhinoceroses from which they remove ticks and insects. The Egyptian plover actually ventures into the open mouths of crocodiles to collect leeches from between their teeth. In the sea over forty species of fish are known to have the habit of cleaning other fish and there may be many more. About six species of shrimp show cleaning activity and also a Galapagos crab which removes ticks from the Marine iguanas, lizards special to these islands. The California Cleaner shrimps *Hippolysmata californica* work in teams and will meticulously clean the surface of fish which in general respond to the presence of the cleaner by becoming docile and cooperative. They remove encrusting animals, parasites and dead tissue. More specialized than this shrimp is the Pederson shrimp *Periclimenes pedersoni* from the West Indies. This shrimp advertises its trade by conspicuous white stripes and violet spots and a special kind of behaviour that involves waving the antennae and swaying the body. When not engaged in its cleaning activity the shrimp shelters in a large Sea anemone. This shrimp works alone and like Cleaner fish has definite cleaning stations on the reef which are visited by fish that come in search of their useful valet. Fish willing to be cleaned will present an appropriate flank to the shrimp which will then clamber about picking off parasites and cleaning wounds. Examination of the stomachs of these cleaner shrimps confirms that they feed on surface parasites.

A fish being cleaned by a Cleaner shrimp which advertises its trade by sporting a distinctive livery of red and white stripes. The fish allows the shrimp to pick off parasites which the shrimp eats so, as in all symbioses, both partners benefit from the association.

Physiological Symbioses. Many invertebrate animals harbour in their tissues bacteria, fungi, protozoans or algae that provide them with useful substances such as food and vitamins. In many cases the association is necessary to the animal and without their microflora or fauna they would die. In return, the partner organisms obtain shelter and may use host metabolites or waste products such as carbon dioxide and nitrogenous wastes. This means that these organisms often enter into biochemical exchanges with their hosts of a rather subtle kind and the details of these kinds of relationships are only recently becoming understood, thanks to the use of radioactive tracers in tagging molecules exchanged between the two partners and due to experiments in which hosts are artificially deprived of their microflora and fauna and the effects of the deprivation studied. Since both partners benefit in these relationships they can be said to be symbiotic or mutualistic. One of the best known examples of this type of symbiosis is the association of wood eating termites with flagellate protozoans harboured in the hind gut. The flagellate protozoans produce cellulose digesting enzymes which break down the wood eaten by termites. Without flagellates the termites would starve to death because, like many plant feeding animals they can produce no cellulase enzymes of their own. So closely integrated are the symbionts that they respond to insect hormones and when the insect moults the increase in level of moulting hormone in the tissues causes the flagellates to reproduce sexually. This results in formation of resistant cysts which can survive outside the host after the parasites have been ejected due to the cuticle lining the gut being shed. This allows the parasites to survive outside their host until they can reinfect another. As we shall see later, there are several examples of parasites too taking clues from host hormones. Bacteria (and perhaps protozoan ciliates) in the rumen of cattle also aid in cellulose digestion.

Some of the strangest kinds of symbiosis are those where an animal harbours luminescent bacteria in its tissues for use as lanterns. This occurs notably in certain abyssal squid and fish. The hosts often have special pockets of tissue to contain the bacteria and these may be well supplied with blood. Although the bacteria emit light continuously due to chemical process requiring oxygen, the squid or fish may incorporate a pigment screen into the light organ to allow intermittent light flashing for signalling or to allow the animal to hide. Even single celled protozoans may contain symbionts. The ciliate *Paramecium aurelia* contains bacteria of different kinds which are found only in certain genetic strains of host. As has been mentioned formation of so-called mate killer substance is due to toxins produced by bacteria (called mu) which

Adult flatworms *Convoluta roscoffensis* owe their green colour to algae symbionts living in their tissues. The algae photosynthesize to provide sugars and other materials for the worms which come out of the sand and lie exposed to light when the tide is out.

kill incompatible *Paramecia* during reproduction. Another species, *Paramecium bursaria* can contain green algae and this too is genetically determined, only certain strains of *P. bursaria* being able to establish a relationship with the algae. The relationships of invertebrate animals with algae are amongst the most fascinating of symbiotic relationships. Many sponges, cnidarians, flatworms and molluscs harbour single celled algae in their tissues. Freshwater animals contain green algae called zoochlorellae and marine animals contain yellow or brown algae called zooxanthellae. Although some algae can apparently exist in a free state they are characteristically found in the tissues of particular invertebrate hosts. The widespread freshwater polyp *Chlorohydra viridissima* owes its colour to zoochlorellae in the cells lining its 'gut' cavity. These are transmitted in the egg. The alga uses host wastes (carbon dioxide, nitrogenous compounds and phosphates) and makes its own food by photosynthesis. Excess soluble food is passed on to the host as well as oxygen, a by-product of photosynthesis. There is some evidence that *Chlorohydra* will resist starving longer than species of *Hydra* without

algae, providing it is kept in the light, so presumably the food provided by the alga is not insignificant. As mentioned earlier, corals contain zooxanthellae which limit the depth at which the corals can grow, since the algal partners need light for photosynthesis. The symbionts may provide food materials for the corals and also help in skeleton formation by absorbing excess carbon dioxide. The beaches of Brittany sport a tiny green flatworm called *Convoluta roscoffensis* which owes its colour to a dinoflagellate algae *Amphidinium carterae* in its tissues. This flatworm comes to the surface of the sand when the tide is out and occurs in such great numbers that it may give the sand a greenish tinge. Possibly this behaviour exposes the algae to sunlight which they need for photosynthesis. The young flatworms have a rudimentary gut and become infected by eating their egg case when they hatch. The algae quickly invade all the tissues of the worm. After a while the gut degenerates and *Convoluta*, no longer able to feed, digests some of its guests. After the flatworm has laid its eggs it gradually exhausts its supply of algae because it eats them at a greater rate than the alga can reproduce inside the host's tissues. The algae live actually inside the cells of the worm so the relationship is very intimate. They have become highly modified and have lost the outer cell wall or theca, the locomotory flagellum and the eyespot. Another flatworm called *Amphiscolops* is unable to reach sexual maturity if deprived of its symbionts so the algae may produce some essential vitamin or other substance necessary for full development of its host. This flatworm is often found in muddy sand in conditions where the oxygen concentration is low and it has been suggested that under these circumstances the algae might be very important in providing oxygen to satisfy the host's normal respiratory needs.

A remarkable example of a large animal that has become dependent on the algae in its tissues is the Giant clam *Tridacna*. This clam can reach a length of 6 ft (1·8 m) and may weigh up to four cwt (200 kg) and it occurs on coral reefs in the Indo-Pacific. Inside the lips of the fluted shell valves the soft tissues are magnificently patterned with splashes of moss green and peacock blue predominating. In these soft lips of tissue and in the siphons the clam farms algae in a highly sophisticated way. The algae live in clusters inside special cells and the clam in effect controls the light intensity to which the algae are exposed by using special lenses to focus light on to the algae or, when the light is too intense, drawing across pigment screens that will protect the symbionts. The algae are eventually digested by special cells in the clam's blood called

phagocytes and the kidneys of the mollusc are specially adapted to deal with the waste products that accumulate in the blood as a result. The clam does not depend entirely on its algal side salad and can still filter feed on organisms collected from the water, but has reduced the size of its gills so that filter feeding is obviously not as important as previously. Work using radioactive carbon (C^{14}) has shown that sugars made by the algal symbionts in photosynthesis are used by the clam to form the mucous crystalline style present in its stomach and used in conventional feeding.

Some Sea slugs store not algae but those parts of algae called chloroplasts which are the sites of photosynthesis. They obtain these from the sea weeds on which they feed. There is evidence that the chloroplasts survive the digestive activities of their hosts and enter the cells lining the dorsal gut processes in a fully functional state. Not only does the presence of these algae help to camouflage the soft bodied Sea slugs; there is also evidence that the algae do produce food substances that the Sea slugs use. Some sugars produced by the algae are known to appear in the mucus of Sea slugs, although it is unlikely that the free chloroplasts divide inside the host tissue. The fact that free chloroplasts can be stored in alien host tissues lends some credibility to the much discussed possibility that the chloroplasts found in multicellular plants have been derived from 'algal' symbionts that originally took up residence in the tissues of their ancestors.

A Giant clam *Tridacna maxima* among corals on the Great Barrier Reef of Australia. The mantle lips of the clam contain symbiotic algae which help to feed this enormous mollusc. The algae are farmed by the clam in a remarkable way.

Parasites. Given the fact that organisms tend to associate with one another for their own ends, it is hardly surprising that parasitism has originated independently in almost every group in the animal kingdom. The echinoderms are one of the few major phyla that seem not to have been attracted to a parasitic way of life: nearly all other phyla have parasitic representatives. As has already been indicated, any of the finely balanced relationships so far mentioned, phoresy, commensalism or symbiosis, could be exploited by one partner which would become a parasite if it constantly harmed its host. Phoretic associations between barnacles, fish and certain invertebrates have almost certainly led to parasitism several times. The Pea crab, *Pinnotheres ostreum*, that injures the gills of oysters and the Mediterranean pearlfish that eats the gonads of its Sea cucumber partner are examples of commensal relationships becoming parasitic. Symbiotic relationships can be equally precarious and there are many examples of fungi which live around the roots of plants at first being helpful to them and then invading them and becoming parasites. Another way parasitism may have evolved is from a predator prey relationship. In cases where the predator does not kill the prey but repeatedly browses off it, the predator may now in some instances be said to be a parasite. This applies for instance to leeches. Originally leeches probably fed on small invertebrates such as snails and worms which they killed; many leeches are non-parasitic and still do this. Leeches that came to feed on larger animals such as fish, birds and mammals, however, did not kill their prey but only sucked blood and so are termed temporary parasites. The dividing line between a predator and a parasite is often a very fine one, especially in the cases of badly adjusted parasites that kill their hosts (though not entirely by eating them, it must be conceded). These definitions also become rather irrational when applied to associations between animals and plants. Most herbivores do not completely destroy the plants they feed on so that the plants continue to form new growth but, nevertheless, we do not term rabbits and cows parasites of grass. An insect that forms a gall on a small part of a plant such as an apple that man has commercial designs on however is definitely given the label of plant parasite.

Parasite adaptations. A lot of nonsense is talked about parasites. They are assumed to be degenerates living unchallenged in the lap of luxury and ease. The truth is that they number amongst the most highly specialized of living organisms, are incredibly finely attuned to the biology of their hosts and have to make constant evolutionary adjustments to prevent themselves from ever losing contact with the host (often a single species of organism) that constitutes the only environment in which they can survive. As we shall see, parasites do not have an easy life; those in warm blooded vertebrate animals for instance, are living in contact with an immune defence system whose one object is to eliminate invading organisms. Invertebrates too have defences against invaders and try to digest or encapsulate them. Mussels and oysters, for instance, lay down pearly nacre around invaders and encapsulate them inside pearls.

Parasites can live either inside (endoparasites) or on the outside of their host (ectoparasites). In either situation they need efficient attachment organs in order to hang on and they tend to lose limbs and other locomotory organs. Fish flukes called monogeneans that live on the skin or gills of fish are good examples of ectoparasites. They are flat and streamlined. They have a sucker or suckers at the back end of the body and these can be armed with an amazingly complex array of spines and hooks. There is another sucker around the mouth and sticky glands on each side of the para-

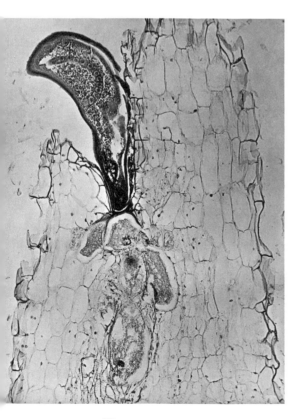

Longitudinal section through a potato root showing a female Potato root eelworm *Heterodera* embedded by the head and about to penetrate and feed in the root.

Looking down on a Giant clam *Tridacna maxima* we see the colourful mantle patterned with pigment and containing algal symbionts meeting in the midline between the margins of the shell.

site's head so that the worms can attach their head ends during feeding, or in order to loop along. Gill parasitic monogeneans have a posterior attachment organ that is subdivided into many clamps. Each clamp usually holds on to a single gill lamella of the host. These parasites cling on so efficiently it is extremely difficult to prize them loose even with dissecting instruments. Obviously the parasites have to expend a certain amount of energy just to remain attached but the attachment organs are cunningly constructed so that only a minimum amount of muscular power has to be used. Some ectoparasites like *Rhizolepas*, a relation of the barnacles parasitic on bristleworms, actually anchor themselves by means of rootlet-like processes that at the same time absorb nutrients. A similar tendency is shown by copepod (shrimp-like) parasites which develop branching head processes that extend into the hosts' tissues. One such parasite, *Lernaea branchialis*, is a common parasite of cod gills and is almost unrecognizable as a crustacean. The head and thorax have become modified into branched processes and only the curled abdomen and egg sacs protrude outside the body of the fish, looking more like part of a worm than a shrimp. Other arthropod parasites like ticks, fleas and lice have claws, bristles and suckers to anchor themselves to their hosts.

Internal parasites can be roughly divided into gut parasites and tissue parasites. Both kinds may need attachment organs and different parasites have different kinds of attachment organ. Flukes have two ventral suckers, one around the mouth and one further back along the body. They also have spines in the body wall. Tapeworms have an organ called a scolex at the head end since there is no mouth or gut, and this is armed with suckers and often with a crown of hooks. Microscopic spines called microtriches on the body wall may also act as an attachment device. Tapeworms do not necessarily remain still inside their hosts but may move up and down the gut following the host's last meal along. Some roundworms such as the hookworm of man, which attaches to the intestinal wall and causes serious blood loss, uses its mouth as an attachment organ. This is lined with cuticle and contains sharp, cutting plates. It must be remembered that some parasite larvae may also have attachment mechanisms, particularly those that penetrate into the host via the skin. Parasites living inside their hosts are particularly susceptible to attack by host antibodies, phagocytes or other defence systems and there is evidence that one parasite at least can trick its host so that the host does not suspect the presence of an invader. This occurs in the human Blood fluke

Schistosoma which lives in pairs (the male grasping the female) in the veins around the gut or bladder of man. As these worms are actually in the blood system they are especially exposed to the host immune mechanisms. The worms are now known to coat themselves with host antigens (large molecules of host origin) so that they remain masked and unrecognized by the host. Some parasites also survive inside cysts in host tissue, which presumably isolate them from harm. *Paragonimus*, the human Lung fluke, does this and so does the larva of a small nematode that lives in the muscle of man and pig called *Trichinella*. Gut parasites like tapeworms, have to prevent themselves from being digested by host enzymes and coat themselves with protective mucus. Internal parasites may have to be able to live in conditions of very low oxygen concentration and many are able to survive with none at all. Parasites also have the problem that if they live in two or more hosts of different kinds during their life cycles, their physiology may have to be adapted to the completely different conditions encountered in each host. *Trypanosoma brucei* is a flagellate protozoan causing sleeping sickness in wild game in Africa and is transmitted by Tsetse flies *Glossina*. In the fly it occurs in the gut and in the warm blooded vertebrate in blood and nervous tissue. The form in the insect is quite different in appearance and respires in a completely different way from that in the vertebrate blood. Parasites that have developed either outside the host, or in a cold blooded invertebrate intermediate host such as a snail or an arthropod, and then enter a bird or mammal with body temperatures of 100–98°F (40–37°C) respectively may have body fats with completely different melting points and use enzymes which work best at different temperatures in their new warm surroundings. The change in temperature involved is in fact often used as a signal to tell the parasite it has arrived in its new host and may serve as a stimulus for further development. Once established in its host in the correct site the parasite has to carry out all its activities without doing too much damage to the host. If it kills the host the parasite is, as it were, committing suicide. Parasites living in the tissues may have difficulties getting eggs or larvae to the exterior if not in connection with natural openings such as windpipe, reproductive and urinary openings or gut. If the parasite is in the gut, the eggs and larvae may have to be protected against host digestive enzymes with shells, cysts or mucus coats. These will also help to protect the parasite larvae in the outside world if they have to have a free living phase between hosts. Parasites that live deep in the tissues or in the blood system have great difficulty getting their larvae out of the host for dispersal to new hosts. The human Blood fluke *Schistosoma mansoni* has evolved eggs with a single sharp spine which help them to bore through the walls of the blood vessels and into the gut so that they can escape in the faeces. The larvae inside the eggs also produce enzymes that help the eggs work their way through. This causes an enormous amount of damage to the gut wall which becomes

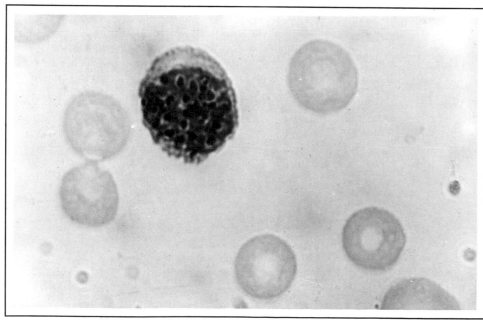

Photomicrograph of red blood cells one of which is infected with malaria parasites. The infected blood cell (dark contents) contains many new parasites and these burst out of red cells all over the body all at the same time this causing the fever typical of the disease.

badly inflamed so the disease schistosomiasis is produced largely by the eggs rather than by the adult parasites. Other blood and tissue parasites such as the Malaria parasite *Plasmodium* and nematode filaria worms use blood sucking mosquitoes as a means both of leaving one host and being dispersed to the next. Adult filaria worms *Wuchereria* live in the lymph ducts of the limbs of man and other animals and can cause the limbs to become grossly swollen, a condition called elephantiasis. In order to transmit offspring, the female filaria worm produces millions of minute larvae (microfilaria) which enter the blood stream. These larvae can be easily seen in blood smears and are one way of diagnosing the disease. The larva occur in the general blood circulation only at certain times (usually at night) so blood smears taken during the day may show no larvae, even though the patient may be badly infected with the filaria parasites. It seems that the larvae appear in the blood only at the times of day that mosquitoes, which transmit the disease, are likely to bite man. At other times the microfilariae remain around the lungs. The behaviour of the parasite is obviously beautifully adapted to match that of its carrying agent or vector, the mosquito, but how does the

Diagram illustrating the life cycle of the human blood fluke *Schistosoma haemotobium* which causes damage to the bladder wall as its eggs bore through to get out of man. The parasite uses a water snail in which to complete its larval life and multiplies enormously once inside the snail.

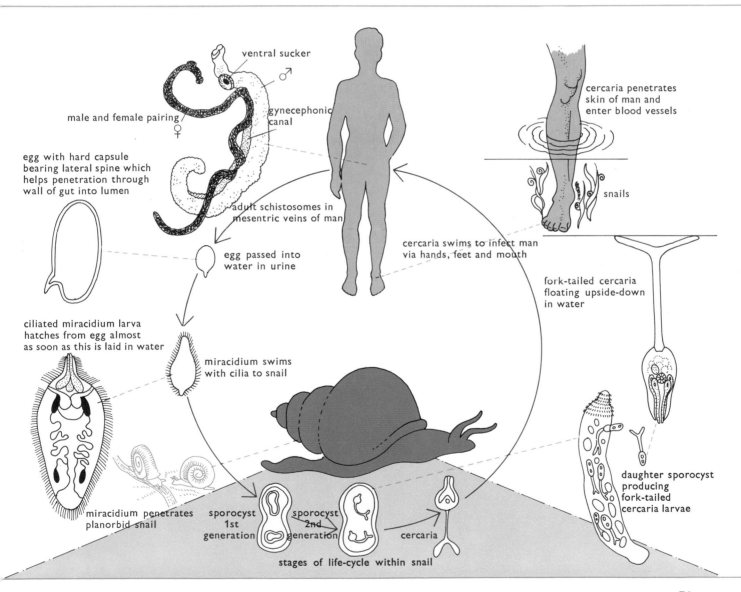

parasite set its clock to know what time of day it is outside the body of man? The answer to this seems to be that the parasites respond to changes in the oxygen concentration in the blood which occur diurnally due to man becoming active or resting. If a host reverses his activity phases and becomes active at night and sleeps during the day, then the parasite rhythm also reverses and filaria larvae now appear during the day. There is some evidence that malaria parasites also have a clock. The fact that different kinds of malaria parasite cause fevers at either 48 or 72 hour intervals immediately suggests that each species of parasite is in synchrony because the fever results from the malaria parasites all bursting out of red blood corpuscles at the same time and releasing toxins. The malaria parasites are all reproducing at the same rate all over the body. The forms of malaria parasite that burst out of the red corpuscles during bouts of fever cannot

infect mosquitoes, only more host red blood cells, but at the time that these (merozoites) burst out, forms which are able to infect the mosquito (gametocytes) are just starting to form and will mature roughly the night after the day the fever occurred. This means that because the gametocytes start their development at the right time they are most abundant at night, which is when most species of mosquito feed. The enormous adaptability of parasites is shown by the fact that on the Pacific Islands there are mosquitoes that bite during the day. The Pacific strains of filaria worm have adapted to this because their microfilaria larvae appear in the bloodstream during the day instead of at night, as elsewhere in the world. Some tissue parasites have to remain in the host tissues until the host dies and is eaten. This occurs in, for instance, the bladder worms of the Pork tapeworm *Taenia solium* and *Trichinella*, a nematode, both of which

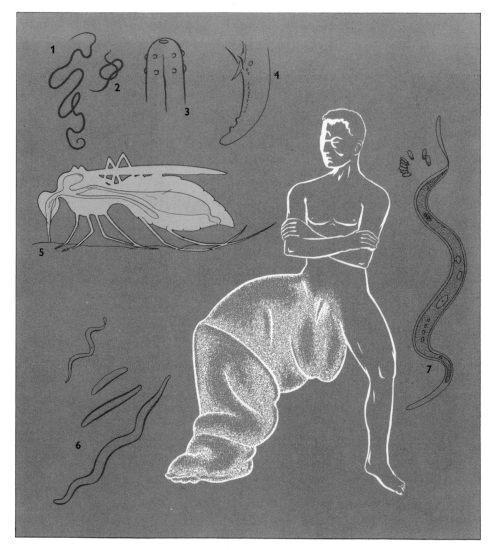

A victim of elephantiasis with the various stages in the life cycle of *Wuchereria bancrofti*, the parasitic roundworm that causes this condition. The female (1) and male worms (2) are here drawn life size. The females have papillae (3) and the males a spine (4) with which they attach themselves in the lymph ducts of man, causing swelling. They produce larvae called microfilariae (7) which enter the blood and are sucked up by mosquitoes (5) when they feed. Inside the mosquito, the larvae develop into infective stages (6 and 7) which are injected into the blood of another individual when the mosquito feeds again.

can be transmitted in undercooked pork. A few parasites unable to escape from intermediate hosts use every guile to make the host visible to predators that form the final host. The larval stage of one fluke *Distomum macrostomum* that parasitises a snail *Succinea* and a thrush, turns the tentacles of the snail into banded, pulsating beacons so that it becomes very conspicuous to the thrush which then becomes infected. It is the striped football-sock like body of the parasite visible through the thin wall of the tentacles that causes this transformation.

Because of the difficulty parasites experience in locating new hosts, large numbers of eggs and infective larvae have to be produced to give some a chance of reaching their goal. Many parasites are hermaphrodites that is, both sexes are present in one individual, and this is not necessarily to allow self fertilization to occur. Self fertilization is in fact a rare occurrence even in parasites. Parasites are hermaphrodite so that all individuals can lay eggs, not just the females, and this increases output enormously. Some parasites have a phase of larval multiplication which further increases their numbers. This happens in digenean flukes such as the Liver fluke of sheep and cattle and the human Blood fluke, both of which use snail hosts in which to multiply. To ensure that they get into the right host at the right time parasites may have to gear their reproductive activities to those of their host. Several parasites reproduce at the same time and develop at the same rate as their hosts so that the parasite offspring will be around at the same time as the susceptible young host animals. In order to achieve this synchrony the parasite may take clues from host reproductive hormones which also cause the parasites to start reproducing. This happens in the frog parasites *Opalina* and *Balantidium* (protozoans in the rectum) and in *Polystoma* (a monogenean fluke in the bladder). Both are activated to reproduce by host reproductive hormones. The breeding cycle of the Rabbit flea is also controlled by the reproductive hormones of its host. Fleas only become mature and mate after having imbibed blood either from a pregnant doe or from a young rabbit 1–7 days old. The pregnancy hormones of the rabbit are sucked up in the blood meal and exert a physiological effect on the fleas.

The larvae of parasites have two main functions, dispersal and host location, and they are extremely well adapted to perform these functions. Broadly speaking there are two main kinds of parasite larvae, those that remain inside egg shells, sheaths or cysts and are swallowed passively by the host and only become active once they are inside the host and receive the appropriate chemical and temperature stimuli, and those larvae that are free living and

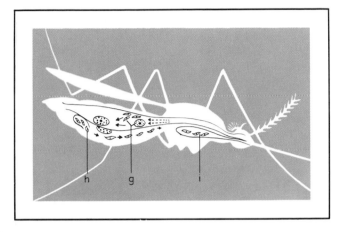

Malaria parasites are transmitted by *Anopheles* mosquitoes. When the mosquito sucks blood, sex cells or gametocytes are taken into the insect's stomach. The gametocytes seem to be produced in the blood only at the times of day that mosquitoes bite showing how beautifully the parasite is synchronized with its host.

active, often with well developed sense organs and nervous system, that have the onus of actually locating the correct species of host and penetrating it. Parasites that locate their hosts actively are known to use chemical clues such as fatty acids in mucus, chloresterol in sweat, carbon dioxide, temperature and vibration. A particular larva has a built in programme of 'expected' stimuli that have to be satisfied in the right order before the next behavioural phase, or further development, leading to colonization of the host will occur. These stimuli that help parasites recognize their host act as switch mechanisms for the unfolding of a complex behavioural programme rather like the homing of a missile. These homing mechanisms must be incredibly precise because parasite larvae are able to locate the correct species of host from other closely related species that occur nearby in the same environment.

This outlines a few of the many complications that parasites encounter in their chosen but not necessarily easy way of life, and some of the brilliant solutions that have been evolved to overcome these apparent obstacles to their success.

Sense Organs and Behaviour

The world we live in is very much a product of our sense organs. If we had eyes that could perceive only ultraviolet light, for instance, or ears that, like those of bats were sensitive to very high frequencies, we should probably interpret our surroundings very differently. In the end it is anyway the brain that sees and hears using the information fed to it from the sense organs. It is impossible to know how invertebrates experience their world merely by investigating the modalities of their various sense organs, although this gives us clues. This is because the process of experiencing is brought about by the integration of information not only from different sense organs but also from the integration of sensory input gained over the whole past history of the animal (and indeed perhaps over a period of evolutionary history if innate 'experience' is involved). There is a sense in which animals *learn* to see, taste, smell and so on, though animals that use innate experience do come into the world very highly programmed to switch on certain patterns of behaviour when the appropriate kind of sensory information is fed in. Each animal inhabits a private subjective world and this is not only because of the

kinds of sense organs it has and the way that sensory information is processed but because animals perceive only biologically *relevant* stimuli, that is information essential to their success. Nocturnal or deep sea animals tend to be colour blind because colour has become irrelevant to them in their dim surroundings; animals that inhabit perpetual darkness such as many cave living vertebrates and internal adult parasites have lost their eyes because vision has become of no use. Parasitic animals are interesting in this way because, although the adults may have lost their eyes, these are often present in larval stages responsible for dispersal of the species and host location. The same is true for many attached, filter feeding invertebrates such as mussels, barnacles and Sea squirts. These all have pigmented eyespots as larvae but lose them on becoming sedentary adults, although some bottom living invertebrates may still be sensitive to diffuse light.

The ability to respond to environmental stimuli seems to be a basic property of living matter. Even viruses, which consist of little more than a protein coat and a strand of nucleic acid, can respond to host cells by penetrating into them. *Amoeba*, a single celled protozoan, shows well developed chemical responses and can identify food and avoid acid or alkaline materials using this sensitivity. Similar kinds of sensitivity are shown by mammalian white blood cells that engulf foreign bacteria and other invaders and by cells of developing embryos that communicate and interact during tissue and organ formation. Some flagellate protozoans have an eyespot or eyespots plus a shading device associated with the whip-like flagellum which is a locomotory organ. Stimulation of the eyespot directly affects the movements of the flagellum and this brings the protozoan into conditions of optimum light intensity. As many of these flagellates are plant-like and use the sun's energy to make their own food from simple substances by photosynthesis (for instance *Euglena* and *Chlamydomonas*) the ability to locate areas of high light intensity is of considerable importance.

The most 'primitive' invertebrates to have a nervous system and sense organs are the Sea ane-

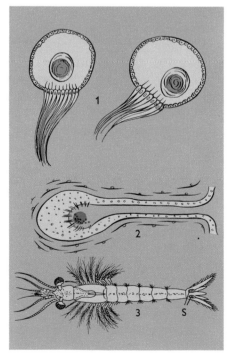

Statocysts are particularly important in animals that float or swim. (1) a typical statocyst with its chalky granule (pink) which touches sensory nerve endings when movements of the animal's body cause it to roll around (left). (Right) nerve fibres from the statocyst run to the brain. (2) Statocyst of a Common mussel, and (3) an Opossum shrimp *Mysis* with statocysts in the tail.

Cushion starfish like these Duck's foot starfishes can turn themselves over by filling parts of their body with water.

mones, jellyfish, *Hydra* and allies, these being the stinging animals or cnidarians. Sponges, it will be remembered, have no sense organs nor a nervous system, or at least one has not yet been found. They do, however, show a certain amount of co-ordinated behaviour and can contract the exhalent pore or pores through which water is expelled from the body. This may take as long as ten minutes after stimulation by prodding the sponge. Presumably the ability to close up is useful in an intertidal animal which needs to conserve water when the tide is out. Complete contraction of a sponge may sometimes be induced by repeated poking and the stimulus may produce an effect in regions several centimetres away. Not only is the mechanism of contraction not understood (sponges have no muscles similar to those of higher animals), the way that the stimulus is transmitted and the co-ordinating activity produced without a nervous system is also unknown. Sponges are not always as well co-ordinated, however, and sometimes the exhalent opening can be induced to close without the flagellate cells inside the sponge seeming to be aware of this so that they continue beating. This has the unfortunate consequence that water continues to enter the sponge through the inhalent openings until the sponge pumps itself up and bursts!

Getting Around Without a Brain. Compared with even simple metazoan animals like Sea anemones and jellyfish, the sponges behave as a colony of individual cells rather than as an integrated organism. The Sea anemones, corals, jellyfish and *Hydra* have no brain but do have a network of small nerve cells straggling between the two main cell layers of their bodies. This nerve net acts as a co-ordinating system linking sense organs with muscles and other effectors such as the stinging cells. At one time it was thought that stinging cells were individually sensitive to the appropriate stimulus such as touch or chemical action and that the flagellar trigger at the side of the stinging cell would automatically cause ejection of the nematocyst thread on reception of the stimulus. Now it is known that these so-called independent effectors (because they both receive impulses and act as an effector) are linked into batteries by the nerves and that their level of sensitivity can be set so that they do not become discharged needlessly. Presumably the relationship of certain corals to fish, like the Clown fish of the genus *Amphiprion* that forms an association with them relies on the fact that the coral somehow 'learns' to withold discharge of the stinging cells for that particular species of fish even though corals are carnivores and may eat other small fish. Sea anemones that live in association with Hermit crabs show very complex behaviour patterns and play an active part in locating occupied whelk shells and crawl on to them by attaching the tentacles first and then hauling up the basal disc.

These climbing movements are rather like the co-ordinated somersaulting or looping mode of locomotion shown by *Hydra*. The anemones must be very much 'in touch' with the crab because when the Hermit crab moves out to find a larger shell, the anemone soon follows suit. Free living anemones like the Beadlet anemone may be highly aggressive competing for space and holding territories. This they do by stinging and damaging members of the same species with which they come into contact. All this suggests that, brainless though they may appear and in fact be, Sea anemones are fairly sophisticated.

Some jellyfish have quite well developed nerve rings around the outside of the swimming bell, an ideal communication link for a circular animal. Eyes are present in some jellyfish and are arranged around the outside of the bell. Gravity receptors, called statocysts, are also situated peripherally and act as balancing organs so that the jellyfish can control the position of its body in the water. The statocysts of jellyfish are very similar in principle to the balance organs that humans have in their inner ear. Basically a statocyst consists of a fluid filled sac containing a chalky granule that rolls around on a bed of ciliated or flagellate sensory endings and stimulates them in different ways according to the way that the granule impinges on

them due to gravity and rotational body movements. Statocysts occur in many invertebrates as well as in vertebrates. They are especially important in marine invertebrates which swim or float suspended in a fluid medium and which have to be able to detect changes in course, but they also occur in land animals such as the Garden snail, *Helix*. Other animals with statocysts are some flatworms, *Octopus*, scallops, mussels, lobsters, some tubeworms and Sea cucumbers. The balance organs of *Octopus* are comparable in complexity to the vertebrate inner ear. In addition to the statocyst, there are three canals at right angles to one another which detect angular accelerations. Removal of these balance organs from the cartilage pockets on each side of the brain case causes the animal to become completely disorganized when swimming and to trip over its own arms when walking.

Echinoderms. Like the stinging animals the echinoderms, which are the Sea stars, Sea urchins, Sea cucumbers and allies, are organized on a circular or radially symmetrical plan. As there is no head there is no brain, merely a nerve ring running around the mouth giving off nerves which run along the arms in a Sea star or along the grooves between the tube feet in Sea urchins and Sea cucumbers. Nevertheless, Sea stars use this simple nervous system to advantage and are effi-

◁ **Many animals like this European lobster can be hypnotized, usually by stroking between the eyes, a much more humane way of capture than with a harpoon gun!**

A starfish rights itself after having been turned upside down by ▷ **curling the arms on one side under its body.**

Mediterranean rock boring Sea urchin *Paracentrotus lividus* uses its spines to make a niche for itself. Stones and shells may be used as 'sunshades' in the morning and are held in position on the spines by the tube feet.

cient predators. They have a rudimentary eye at the tip of each arm and the general body surface may also be sensitive to light. They also have a well developed sense of smell and will trail food moved around an aquarium, if hungry. Their appearance in lobster pots is doubtless associated with their ability to scent out the bait. Sea stars are also sensitive to touch and apparently to gravity because starfish can turn over if placed upside down by curving the arms on one side under the body whilst detaching the others, then somersaulting over. Pentagonal Sea stars do this by filling different regions of the body with water to tilt the body over. Apparently Sea stars can learn, and do have a memory because a Sea star taught to turn over using one particular arm retained the memory for five days without further conditioning. On another occasion Sea stars taught to draw back at a

boundary between rough and smooth and dark and light surfaces retained the memory for 2–3 months without further training (obviously star pupils). It has also been found from these kinds of experiments that if the nerve ring of a trained Sea star is damaged the memory may be regenerated along with the new portion of the nerve ring, although understandably there is a period of disorganization.

Sea urchins are in contrast peaceful herbivores which are sometimes preyed upon by starfish and by sharks, rays and other animals. They show interesting defence reactions. Their spines, tube feet and pedicellariae (small protective forceps-like projections from the surface) are all sensitive to touch and may also respond to chemicals, for instance those released by predators into the water. The pedicellariae can sometimes respond to the smell and touch of a predatory Sea star by nipping at the attacker's tube feet by way of discouragement. In Sea urchins with long spines such as the black mediterranean Sea urchin *Diadema* touching the body causes all the spines to point to the region touched. The animal then moves away from the stimulus. Sea urchins also show a generalized light sensitivity all over the body although a structure called a podial pit on the top of the animal may be particularly important. Some urchins daily cover themselves with stones and shells held in position with the tube feet; this occurs a few hours after dawn and may protect light sensitive pigments in the body. Sea urchins may also show a shadow response and make rapid movements with the spines when a shadow suddenly falls across them. This is a protective response also found in other invertebrates as will be described. Space does not permit mention of all those invertebrates that manage without brains; suffice to mention that many sedentary animals attached to a rock, sea weed or the sea bed, for instance, oysters and mussels, have found ways to survive without brains.

Invertebrates with Heads. Animals that move along head first tend to have a co-ordinating brain at the head end because this is where all the sense organs are concentrated, as a kind of early warning system. The sense organs also tend to be arranged in pairs on each side of the body so that the animal can compare stimuli coming from each side. The way that this is useful can be shown by imagining an animal navigating up a concentration gradient of food substances. As soon as the animal swings its head too far out of the path of the food substance to one side it loses the stimulus on this side and corrects course; the same happens if, after correction, the animal veers off course too far in the opposite direction. Comparison of stimuli on each side of the head allows an animal to home on a stimulus source with accuracy. This system also works in reverse for animals trying to avoid a stimulus. Animals that respond only diffusely to a stimulus rather than directionally, however, may have sense organs scattered over the body in a more random way. The earthworm is an example of this and has a number of single celled eyes each containing a lens. These are scattered throughout the skin and concentrated in the prostomium, the front tip of the worm. These are mainly important in enabling the earthworm to avoid light since this causes pathological changes leading eventually to breakdown of the earthworm's tissues; by no means all animals are sun worshippers. The Coat of Mail shells (chitons) also have eyespots called aesthetes scattered over their shell plates. The scallop *Pecten* is exceptional compared with other bivalves. Not only can it leap around, it has two rows of unnerving metallic blue eyes and other sense organs around the opening of the shell. The blue sheen comes from the reflecting layer or tapetum behind each eye. The eyes probably cannot interpret shapes but detect changing patterns of movements particularly light-dark changes moving at speeds similar to those that familiar predators such as whelks and starfish would move at. The tentacles

The jewel like eyes of the scallop probably cannot interpret shapes but detect moving objects. They are also very sensitive to sudden shadowing and signal to the scallop to close its valves quickly.

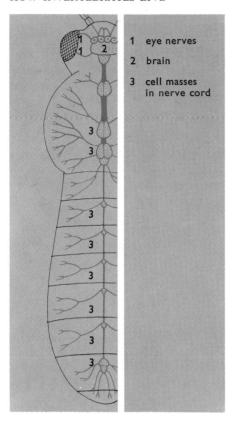

1 eye nerves

2 brain

3 cell masses in nerve cord

Diagram of the nervous system of a midge. Many invertebrates are segmented (for example worms and arthropods) and in addition to the large brain in the head have smaller nerve centres called ganglia that coordinate the activities of each segment.

round the mantle edge are chemo-sensitive and respond to the smell of starfish predators in the water. The cockle is far less prodigious and has small eyes round the top of the siphons only. This forms a kind of periscope for the animal way beneath in the sand and responds to shadows, causing the siphons to be withdrawn rapidly. The shadow response is a protective escape reaction which takes various forms and is also shown by the

scallop, Feather duster worms which rapidly shoot down their tubes causing the crown to be withdrawn and, as mentioned, by the Sea urchin *Diadema*. The shadow response of the scallop which causes the shell to close rapidly has been studied in great detail. Remarkably enough those steely blue eyes of the scallop contain two kinds of retina. One kind responds to light by firing off nerve impulses, the other kind which records the shadow fires off nerve impulses when the light is suddenly cut off! That is, it responds to sudden darkness. These two kinds of retina are quite different in structure when viewed with the electron microscope. The retina that signals the presence of light is made up of light sensitive cells ending in minute protections called microvilli, whereas the cells that signal the onset of darkness have cilia in them. These two kinds of retina correspond to two different kinds of eyes found throughout the animal kingdom. Microvillar 'eyes' are called rhabdomeric and are found in worms, molluscs and arthropods whereas ciliary eyes occur mainly in vertebrate animals including man but occasionally crop up in invertebrates and may in fact be much more common than has been supposed previously. Scallops are not the only kinds of animals to have both kinds of 'eye'. A parasitic flatworm larva that infects Dover soles has recently been found to have two pairs of rhabdomeric eyes surrounded by pigment cups as well as a pair of what are presumably ciliary eyes. Whether the ciliary 'eyes' are shadow receptors is not yet known.

The other way of classifying eyes is on the basis of their overall arrangement. There are two kinds of eye structure in the animal kingdom, compound

One of the paired stalked eyes of the predatory West Indian conch peers out of its shell opening.

Compound eyes of a dragonfly (far right) consist of many tubular eyes called ommatidia. Each of the tiny hexagons is the opening of one of these tubular eyes. Compound eyes are very good for detecting movements and are particularly well developed in predators like the dragonfly.

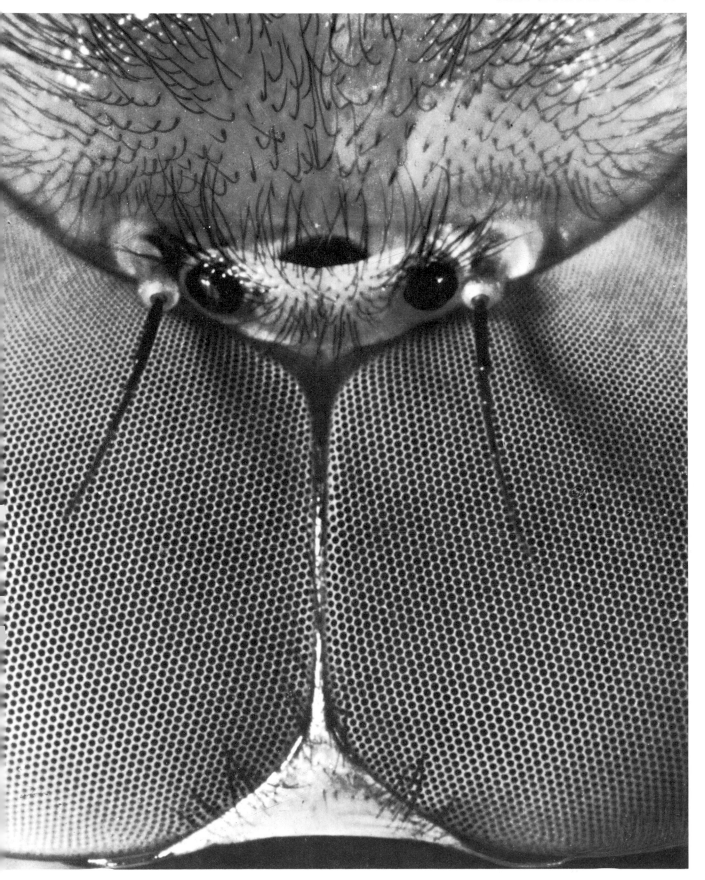

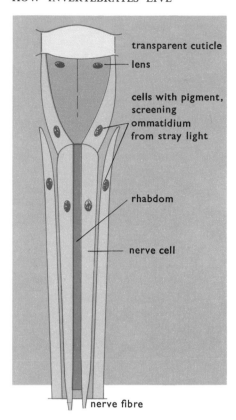

transparent cuticle

lens

cells with pigment, screening ommatidium from stray light

rhabdom

nerve cell

nerve fibre

Diagram of a single ommatidium thousands of which form the compound eye of arthropods and other invertebrates.

eye is usually encased by a cup of pigment cells lying behind the retina. In the compound eye, pigment screens separate each element in bright light but can be withdrawn in dim light so that cross illumination of neighbouring tubular eyes can occur, to maximize light input. Camera eyes have been evolved independently many times in the animal kingdom and occur in flatworms, snails, scallops, some spiders and notably in the octopus, squid and cuttlefish as well as in the vertebrates. The camera eye of the octopus and the other cephalopods is remarkably similar to vertebrate eyes, having a focusing lens, an iris and a retina with screening pigment behind. There are, however, important differences that show this eye has been evolved quite separately from our own eye. The most important of these are firstly that the retina is rhabdomeric in the octopus, ciliary in vertebrates. Secondly that the nerve fibres come off from behind the retina in the octopus but lie in front of the retina in vertebrates.

Simple compound eyes occur in a few marine bristleworms and in many arthropods reaching maximum development in the insects. The main advantage of compound eyes is that they detect movements much more sensitively than single lens eyes, this being of particular advantage to flying or rapidly moving insects.

Other Kinds of Sense Organs. Sense organs are usually classified as photoreceptors (eyes), mech-

eyes made up of many tubular eyes called ommatidia bundled together each ommatidium having a fixed focus lens, and the camera eye with a single lens, often with muscles for focusing situated at the entrance of a cup of retinal tissue. The camera

A Common octopus *Octopus vulgaris* of European seas (far right). ▷ These are highly intelligent invertebrates with large brains and good learning ability. The eight arms have a well developed sense of touch and taste. The eyes are also well developed but are not used as much as in its faster relations the squids and cuttlefish.

◁ In insects, the eye consists of a large number of ommatidia. Each ommatidium perceives only one point of the object; the more ommatidia there are, the clearer the image.

anoreceptors (responding to touch, water currents, gravity, sound), and chemoreceptors (taste, smell, CO_2 concentration etc). There may also be temperature receptors, although in fact most kinds of nerve endings are incidentally temperature sensitive, humidity receptors, osmolarity receptors monitoring for instance the strength of surrounding sea water solution and sense organs sensitive to electric currents (as occur in electric fish – none are yet known in invertebrates). All these mainly keep the animal informed about changes outside the body but there are also sense organs that provide information about conditions inside the body.

These interior receptors are called proprioceptors. Many of these are concerned with telling the animal how stretched its muscles are so that movements can be delicately controlled. The balance organs that tell the body where it is in space are also proprioceptors.

Giant Nerve Fibres. Animals that can make rapid escape movements such as earthworms, tubeworms and fast moving cephalopods have giant nerve fibres running from the brain down most of the length of the body. Sometimes these giant fibres innervate the muscle cells directly so that they form a kind of 'hot line' to the effector organs and allow minimum delay between perception and response. These giant fibres may be absolutely huge. The tube worm *Myxicola* has a giant nerve fibre 0·07 in (1·7 mm) in diameter. The earthworm has three giant fibres in the ventral nerve cord and the largest of these is 160 microns (ie 16/100 of a millimetre) in diameter. This is still giant compared with ver-

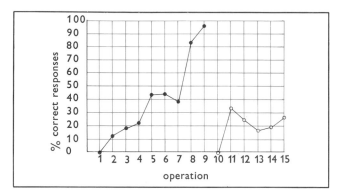

The effect of brain damage on memory in *Octopus*. The animal was trained to attack a crab. After 8 days, part of its brain was removed and it was unable to remember what to do—otherwise it was apparently normal.

tebrate nerves which usually measure less than 20 microns across and have increased conduction rate by the use of fatty sheaths around nerves rather than by increasing the diameter of the nerves themselves. The median giant fibre in the earthworm conducts information from head to tail. The two laterals conduct from tail to head. This makes sure that neither end of the earthworm will be unpre-

pared when confronted with the early bird. The fast moving squids also have giant fibres which run from the brain to the body wall muscles of the mantle. Squids are propelled by jet propulsion produced when the muscular bag of the body wall contracts forcing water out through the siphon. This incidentally produces a respiratory current at the same time because the mantle 'bag' also contains a pair of gills. All the muscles have to contract in unison all the way round the body for this to be efficient. Squids in fact have three sets of giant nerve fibres on each side of the body. Two send processes to the mantle muscles and to the funnel and a third links the giant fibre systems on both sides of the body so that both sides will contract simultaneously. Squids can move extremely rapidly and may reach jet speeds of 20 knots. The octopus which is much slower and tends to shamble around the sea bed rather than shoot overhead has no giant fibres. The enormous size of squid giant fibres which, it must be remembered are processes of single cells, has led to their being extensively investigated by nerve physiologists who frequently spend whole summers at pleasant coastal resorts waiting for the next catch of squid. In fact squid giant axons have proved an extremely useful model for nerve conduction in general and most of the information we now have about the working of vertebrate nerves was elucidated on these molluscs.

Intelligent Invertebrates – the Cephalopods. The squids, cuttlefish and octopuses have large brains enclosed in a cartilaginous brain case rather like our

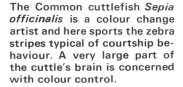

The Common cuttlefish *Sepia officinalis* is a colour change artist and here sports the zebra stripes typical of courtship behaviour. A very large part of the cuttle's brain is concerned with colour control.

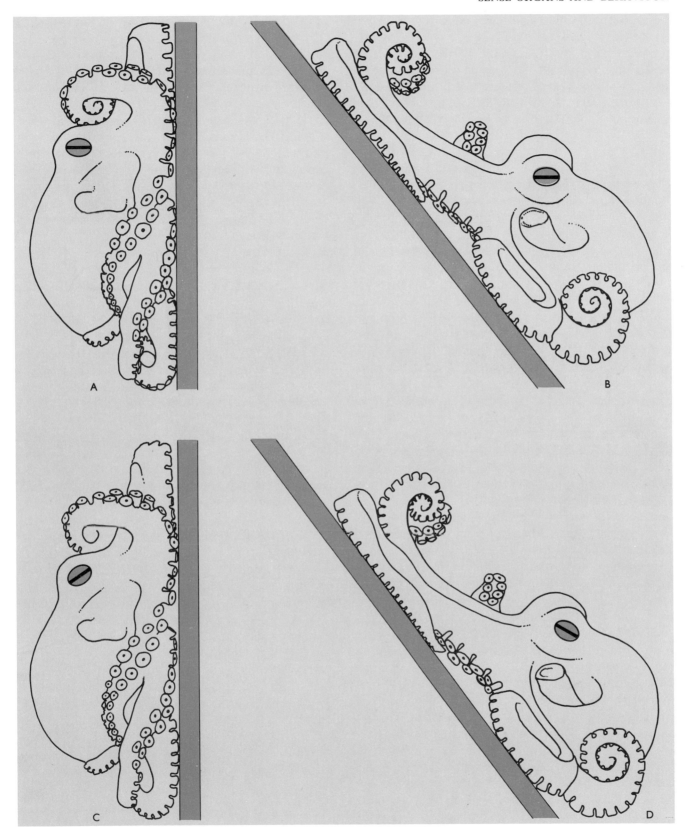

An experiment that demonstrates the functioning of the statocyst in the octopus. Normally the slit-like pupil of the octopus eye remains horizontal whatever the position of the octopus's body (A and B). If the statocysts are removed, the slit changes its angle according to the position adopted by the octopus (C and D).

skull. Balance organs similar to our own inner ear lie at each side of the brain case and on the head are a pair of large eyes again similar to our own. Taste and touch receptors are present on the tentacles. The cephalopods are therefore exceptional amongst molluscs and it is difficult to remember that they belong to the same group as the oysters, limpets and whelks! Cephalopods show complex behaviour for instance in swimming, in catching prey, in distracting predators, in courtship rituals and in the case of the octopus, building homes from cairns of stones. A certain amount of their behaviour seems to be innate. Cuttlefish feed on certain small shrimps almost as soon as they are out of the egg and will squirt ink if disturbed. The octopus seems to have an inbuilt predilection for crabs which it prefers to eat above all else. Despite a certain amount of inherited, innate behaviour, the octopus and cuttlefish can learn new behavioural patterns. Both cuttles and octopuses can be trained to come for food and at a more practical level the octopus can learn not to squirt inky water at visitors to its tank. This does not make the octopus easy to keep. They are born escapologists and will squeeze their bodies paper thin through any crack if determined to escape. The ability of an octopus to show conditioned learning has led to much work on its vision, taste and touch. The octopus can be trained to recognize different shapes by presenting it either with a reward (being allowed to eat a crab) or with a deterrent (an electric shock). After repeated trials the octopus learns to recognize the difference between certain shapes even when these are now presented alone without either food or a shock. Using such experiments it has been found that an octopus can differentiate between squares and diamonds quite easily but, for instance, finds it difficult to distinguish between a square and circle of approximately the same area. Their learning ability can also be tested in mazes with a crab as goal. Octopuses probably have colour vision and indeed this would be expected in an animal that itself has a highly developed colour system in its skin. The skin of the octopus, cuttlefish and, to a lesser extent the squid, contains many thousands of pigment bags called chromatophores. These contain various coloured pigments including white and are arranged in layers in the skin. They have muscular walls which can be contracted to reduce the colour spot to a mere pinpoint or expanded to a large spot. Each chromatophore is operated by nerves so that very rapid shifts in colour pattern can

be produced. A very large part of the brain of these creatures is concerned with colour control. The cuttlefish in particular produces remarkable colour changes. Waves of black ripple over the body if a predator such as a dogfish swims above. Two glaring black eyespots can be produced against a starkly white back in another defence response. Cuttles can also match the background of the sand, into which they partially burrow, almost perfectly and change the texture as well as the colour of the skin by drawing it up into small bumps like goose-flesh. The remarkable zebra stripe display of threatening male cuttles has already been mentioned. Vision is particularly well developed in the cuttles, squids, and the octopus, although visually proficient, relies much more heavily on taste or touch receptors on the tentacles for information. It has been shown that the octopus is 100–1000 times more sensitive to certain chemical stimuli than humans. In addition, the octopus can learn to distinguish between perspex cylinders and other objects grooved in different ways to give subtly different textures, so has in those apparently unwieldy arms a delicate sense of touch!

Colour Control in Crustaceans. In this world of adaptation and counter adaptation it is not just the octopus that changes colour but also its prey, the crabs and their allies (the crustaceans). Animals preyed upon by predators with good vision need a convincing colour camouflage. We have already seen that the octopus changes colour by contracting or expanding the muscular pigment sacs in its skin. Crustaceans like cephalopods have chromatophores in the skin under the cuticle. Each contains a different coloured pigment, white, black, red or yellow and the differently coloured chromatophores are arranged in layers and can produce intermediate shades, for instance brown, in combination. Crustacean chromatophores are controlled in a completely different way from those of cephalopods. Instead of the pigment sac contracting, a substance is secreted that causes the pigment itself to clump. So how does the crustacean know what colour to become, and what causes the colour pigment to become clumped? Basically, a crustacean receives information about its background via the eyes. Information received via the eyes causes the release of neurosecretory hormones that control

Sponges, like these much magnified intertidal forms known ▷ as *Sycon*, have no visible nervous system or muscles similar to those of higher animals yet they make coordinated movements.

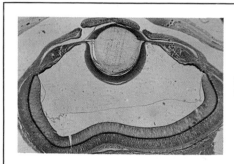

Section of a cuttlefish eye showing its great similarity to the camera eye of vertebrates. The round lens is at the top; the retina is grey.

the expansion and contraction of the various pigments which in turn enables the animal to vary the colour in relation to the background. These hormones or chemical messengers are produced in nerve cells and released into the blood. From here they reach the chromatophores. The system is not as rapid as that of the octopus where pigment cells are controlled directly by the nervous system. The crustacean system involves antagonistic hormones in fact and so a balance is kept between pigment clumping hormone and pigment expanding hormone. Both are usually present in different amounts at any time. The amounts of these two pigments can be delicately controlled by the ratio of direct and reflected light affecting the upper and lower parts of the eye. If either the upper or lower part of the eye is blacked out the animal will be unable to respond appropriately to its background.

Sex and Other Methods of Reproduction

As vertebrate animals we tend to take sex rather for granted as a basic 'fact of life' but in many invertebrates sexual reproduction plays a secondary role to asexual multiplication. Asexual reproduction simply involves division of some part of the animal body into two or more pieces which then develop into whole new individuals. Some invertebrates divide into more or less equal pieces each of which then regenerates into a whole individual, others behave rather like plants and bud off a small region which becomes reorganized into, at first, a miniature individual. Sponges, *Hydra* and Sea squirts can all bud off small copies of themselves and these then grow to normal size. Budding is often associated with the formation of colonies, particularly those where the daughter individuals remain in contact like multiple Siamese twins, often with a considerable area of fused tissue. Sponges, although capable of producing eggs and sperms, often reproduce asexually to form colonies of similar individuals. They do this either by budding of a new individual or by constricting the ends of their branches so that these fall off and regenerate into new individuals. The high regenerative capacity of sponges is employed in the restocking of the overfished grounds of commercial sponges off the Florida coast. Pieces of sponge are weighted down with cement blocks and 'planted out' into the water. These will each grow into marketable sponges after a few years. So great is the regenerative capacity of sponges that a suspension of cells from a sponge that has been dissociated in the laboratory by passing it through a fine mesh will reaggregate into small clumps that soon reorganize into complete sponges. Sponges are frequently coloured, usually in shades of orange, yellow or green. If two differently coloured sponges are dissociated into their component cells and the cell suspensions mixed together, the cells from the same sponge recognize one another and clump together so that two differently coloured sponges are reformed. The cnidarians or stinging animals like *Hydra*, corals and Sea anemones show marked regenerative powers and also reproduce asexually. Sea anemones quite commonly pinch off lobes of the foot which each form small Sea anemones. They may also leave part of the 'foot' behind as they move away and the piece left behind develops into a whole anemone. Corals too bud off new individuals but these remain in contact with the parent and share the same feeding cavity. In the case of Brain corals the new individual is budded off inside the tentacles of the old polyp but does not entirely separate from it at first so that the new mouths remain in contact to form a continuous long trough surrounded by shared tentacles. This gives the surface of the Brain coral its particular convoluted appearance.

Flatworms commonly reproduce asexually as well as sexually. Some small marine (rhabdocoel) platyhelminths divide across the body to form a chain of new individuals which remain attached for a while. The common freshwater planarian *Dugesia* can also reproduce by splitting the body across but the two pieces soon separate by the front half gliding away from the back half which sticks itself down with mucus. Some populations of planarians are only able to reproduce asexually. This is because their cells have an odd number of chromosome sets (eg 3, 5, 7 etc) instead of the usual 2, and are therefore unable to produce viable gametes. Formation of a chain of individuals by rhabdocoels is reminiscent of the asexual budding off of segments (proglottids) by tapeworms. Tapeworms resemble a colony of individuals held together by common muscle, nervous and excretory systems and sharing

Asexual reproduction in *Hydra* with young being budded off from the base of the parent. Later these individuals will break off and grow to full size.

the same head region. Flukes like the Liver fluke of sheep and some tapeworms, such as the Dog tapeworm *Echinococcus*, reproduce asexually in the larval stages which makes up for the low success rate of these larvae in locating the final host. Some marine segmented worms (polychaetes) belonging to the syllids, scrpulids and fanworms (also known as Featherduster worms) reproduce by budding or division. Syllid bristleworms may produce new individuals almost at right angles to the main body which can grow into individuals nearly as large as the parent whilst still attached and then themselves bud, producing a great tangle of interconnected bodies. The earthworm too has a high regenerative capacity and when chopped in half by a spade both portions will often survive to form two whole new individuals. The ability to regenerate in this way depends very much on where the body is divided and is under the control of the nervous system. Molluscs and arthropods tend not to reproduce asexually although some insects may have an asexual stage early on, each egg splitting into several embryos. Sea stars are famous for their ability to regenerate lost arms and maimed starfishes with one or more arms smaller than the others are often

seen. Sea stars often lose arms when preying on bivalves that snap shut suddenly, severing a trapped arm and some Sea stars have made use of their high regenerative powers for reproduction. One remarkable Sea star called *Linckia* can grow a new disc and arms from just one severed arm and sometimes breaks off an arm for apparently no reason, this then growing into a new individual. Other Sea stars commonly split in two spontaneously and each half then regenerates the missing parts. In the old days oyster fishermen concerned about Sea stars attacking the oysters used to drag a dredge over the bottom, collect the starfish, tear them into pieces and throw them back, thereby enabling them to grow into many more oyster hungry starfishes. This is a kind of artificial asexual reproduction, like that produced by chopping earthworms in half, that exploits the natural regenerative abilities of these animals.

One of the single-celled freshwater amoebas, *Amoeba proteus*, is very unusual in that it reproduces only by asexual means either by dividing into two, or by multiple fission where many individuals are produced, each surrounded by a protective cyst. It has sometimes been suggested that because the

The tapeworm from the human intestine (narrow head towards the right) reproduces asexually as an adult by budding off proglottids full of eggs. All parasites need to ensure survival by having a high reproductive rate though this may not be special only to them.

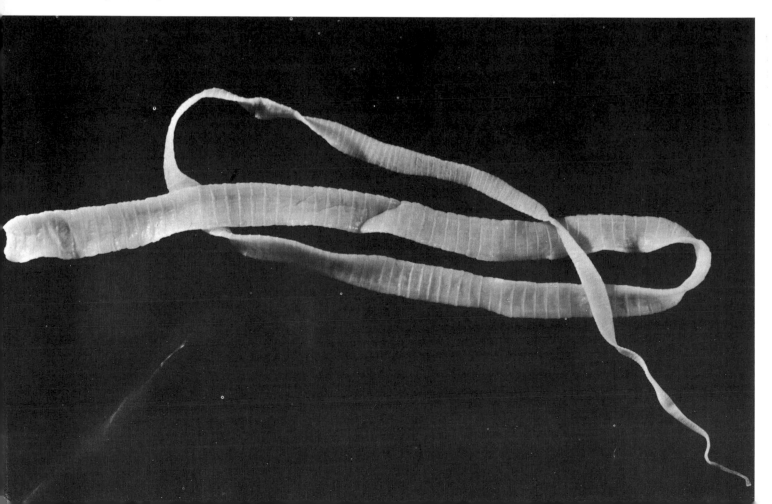

The tapeworm *Echinoccus* has only 3 proglottids or segments. These are budded off from behind the scolex or head as the tapeworm develops in the gut of a vertebrate. The small amounts of asexual reproduction in the adult is more than compensated for in the larvae hydratid cyst stage where millions of new individuals are budded off inside the cyst.

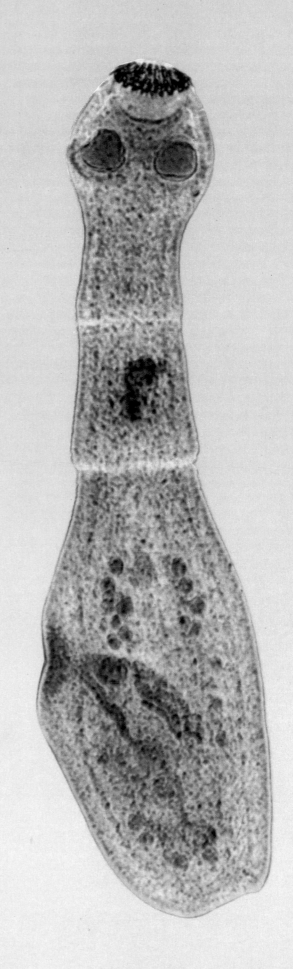

single-celled *Amoeba* divides in half by binary fission it is immortal and is constantly regenerated at each division. There is a sense in which this is true, although it is a delicate philosophical point, but then other sexually reproducing animals are immortal in that their gametes help to propagate new individuals. There is evidence that cultures of the ciliate protozoan *Paramecium* (which also divides by fission but unlike *Amoeba* can also reproduce sexually) do age so that in old cultures the rate of asexual division gradually falls off. It is not certain whether this is due to depletion of some factor in the culture medium or is an intrinsic process. It has been suggested that rejuvenation can occur by a modified kind of sexual process that involves the same kind of nuclear divisions as in the usual sexual process (conjugation) but involves only one individual.

Sex. The whole business of sex has literally been evolved to preserve variety in life. With continuous asexual reproduction one ends up with clones of genetically identical individuals. This may be an advantage in colony formation where fused individuals have to show tissue compatibility, but playing down variability tends to reduce adaptability and leaves animals that reproduce mainly by asexual means vulnerable to environmental change. In sexual reproduction two genetically different individuals each produce gametes by a

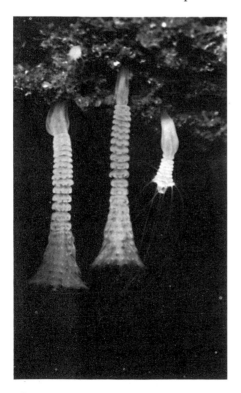

Jellyfish larvae (scyphistoma) split across to form small jellyfish (ephyrae) which swim away from the hanging end. This is a form of asexual reproduction.

process of reduction division (or meiosis) where the chromosome number is halved. During meiosis the chromosome pairs in the gamete-forming cells may exchange portions so that even the process of producing the gametes leads to new combinations of genetic material. Gamete fusion to form a zygote then introduces further variation. By incorporating these kinds of genetic reshuffling process into their life cycles animals maintain enough variability to ensure constant improvement and closer adaptability to the ever changing conditions of life. It is scarcely surprising that most of the invertebrates mentioned as 'being good' at asexual reproduction, above, *also* reproduce by sexual methods. Upon the bare requirement of gamete production and exchange is erected the whole edifice of sex. The technical problems involved are not inconsiderable. Firstly the partners have to be brought together in breeding condition at precisely the right moment. In the case of higher invertebrates they must co-operate and sometimes indulge in complex courtship and copulatory behaviour. They must confine their attentions to members of the same species and so need a set of built in recognition or incompatibility systems which debar sexual contacts with members of different species. Sometimes spawning occurs in which case many partners are involved and a swarming stimulus is transmitted so that mature individuals receive a message to congregate and then to shed their gametes en masse. Very little is known about what controls the precise timing of reproductive maturity and what stimuli are involved in bringing the sexes together in marine invertebrates like jellyfish, Sea stars and Sea squirts.

Sexuality. Animals have been experimenting with sex for some time. Although very often there are two sexes, males and females, which are usually different in appearance (dimorphic), this is by no means always so. Many animals are hermaphrodite, for instance the earthworm, Garden slugs and snails and flatworms. These animals can all produce both eggs and sperm but do not usually produce both at exactly the same time and rarely allow their eggs to be self fertilized by their own sperm. Some animals such as Slipper limpets, as will be mentioned, change sex permanently during their lives, starting usually as males and ending up as females. Others change sex from year to year, oscillating in an undecided way between being a male one year and a female the next. Some animals use sperm only to stimulate division of the egg, not to actually contribute chromosomes to it. This is a situation

Skeleton of brain coral showing the long surface troughs formed by partial division of the polyps which secrete the skeleton.

known as pseudogamy and occurs in a few European flatworms and in roundworms.

Some roundworms, the Water flea *Daphnia* and aphids can spend many generations without having any males in the population at all so that the eggs develop parthenogenetically, by virgin birth. There is usually a restorative sexual phase just before winter to end the virgin summer, however, and the sexual phase is associated with production of an overwintering resistant egg or larva.

Hermaphrodites. Hermaphroditism is quite common in land and freshwater animals that have bisexual relations in the sea and seems in general to have been an adaptation to the more difficult conditions in these habitats. One of the advantages of being a hermaphrodite is that if you meet another mature member of the same species it is bound to be of the opposite sex (and indeed of the same sex) at any given time. This means that not only are the chances of finding a mate doubled but all individuals can produce new offspring, thus doubling the output. Hermaphroditism is particularly advantageous to internal or endoparasites which cannot be sure of meeting another, or many other individuals inside their hosts and need to produce many offspring to offset the difficulties of completing their complex life cycles. As mentioned, few hermaphrodites self-fertilize although some para-

sites such as the Pork tapeworm and the free living scallop and Freshwater mussels do. The Freshwater mussel produces eggs and sperms in different regions of the same gonad. Examples of hermaphrodites other than those already mentioned are some sponges, the freshwater cnidarian *Hydra*, leeches and Sea slugs. The most familiar hermaphrodites, however, are earthworms and Garden snails. These terrestrial animals have the problem of transferring their sperm safely on land and produce copious mucous secretions to prevent dehydration of themselves and their sperm whilst pairing. Most hermaphrodites are acting males first and exchange sperm with a partner. The sperm is then stored in a special part of the female tract of the recipient until mature eggs are produced when sperms are released and fertilize the eggs. Earthworms lie head to tail to pair, usually above ground in the romantic setting of warm dark summer nights. They often keep their tails tucked in their burrows and have only their head ends apposed during the 3–4 hours of mating. Gilbert White, the famous naturalist, noted that earthworms are 'much addicted to venery' and at certain times of year the sight of conjoined worms is not uncommon. The worms hold on to each other by their chaetae and by belts of mucus. Egg laying may begin the day after pairing and can continue for several months without further pairing. The eggs are fertilized as they pass the sperm storage sacs or spermathecae and are laid into a cocoon secreted by a glandular region of skin known as the saddle or clitellum. Several eggs are laid into the cocoon but only one survives as a rule. Some freshwater leeches such as *Hemiclepsis* and *Helobdella* have a rather bizarre way of exchanging sperm. They have no penis but package the sperm into so-called spermatophores 0·2–0·4 in (5–10 mm) long. These sperm packets are placed on a special region of skin on the partner and enzymatic gland secretions on the neck of the spermatophore erode away the tissues so that the sperms enter the partner's body. A kind of string of connective tissue joins the implantation area to the sperm storage regions down which the sperm pass, presumably in response to some chemical secretion although this has not been proved.

Garden snails exchange sperm in spring and summer and have a remarkable courtship that involves the partners firing broadsides of calcareous 'love darts' at one another. These are hard, slightly curved darts about 0·2 in (5 mm) long bearing four longitudinal blades. They are produced by glands associated with the reproductive tract and stored in special muscular sacs. After the powerful stimulation provided by the love darts, mating may last half a day and long packets of sperm are exchanged. Inside the snails the wall of the spermatophore is digested and sperms released into a storage region. About 50–100 eggs are laid a fortnight later in a hole or crevice where they will be protected during their development.

Progressive Sex. Some molluscs have an even stranger sex life than slugs and snails since they actually change sex during their lives, usually starting off as males and then becoming females. The American Quahog clam and the European Slipper limpet (which rejoices in the name *Crepidula fornicata*) are examples of sex changing molluscs. Only 98 per cent of the Quahog population practices sex change, the remaining 2 per cent are staunchly unisexual forms and definitively either male or female. Slipper limpets are quite common around British coasts and live in chains of individuals with the foot of one limpet anchored to the shell of the next. The youngest member at the top of the chain is always a male, individuals immediately beneath this and in the middle of the chain are intersexes whilst the largest members at the base are mature egg-producing females which use the sperm stored from insemination earlier on to fertilize their eggs. It seems that in the Slipper limpet the sexual fate of the youngest member of the association is controlled by the mature females. If these are removed the young male feminizes. This is one way of ensuring that plenty of young males are around for mating since there are no old ones! Sex change may also occur during the life of the ordinary shore limpet since there is a high proportion of males amongst the younger individuals.

Oysters of the kind that brood their young in their gills (for instance European oysters, *Ostrea edulis*) change sex many times in their lives. They all start off as males and, having shed their sperm become females which rely on sperm from young

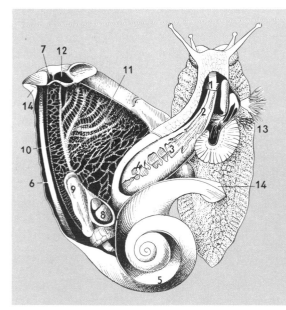

Although land snails are hermaphrodite they copulate to exchange sperm. The snails fire broadsides of love darts at one another as a stimulant during mating. Details of the snail's anatomy are: (1) nerve ring around upper gullet, (2) gullet, (3) stomach, (4) salivary gland, (5) intestinal sac, (6) rectum, (7) anus, (8) heart, (9) kidneys, (10) vas defereus, (11) lung, (12) breathing pore, (13) reproductive organs, (14) point from which the mantle cavity has been cut open.

length acting indirectly via the pineal gland controls the onset of sexual maturity and the timing of sexual cycles in certain mammals. In *Octopus* the importance of light in governing the onset of sexual maturity is shown by the fact that the glands controlling the development of the testes and ovaries in males and females are situated on the eye stalks. It seems a common thing in both marine invertebrates and some insects that once a few members of the population have been brought to maturity by a kind of inner biological clock that is synchronized by light and other external conditions, these mature individuals start to secrete substances that accelerate maturity in other members of the population. These substances, called pheromones, are a kind of external chemical control mechanism frequently acting as a kind of social hormone controlling, and in this case synchronizing, the activities of the group. In animals such as oysters, Sea stars and bristleworms that spawn, further chemical co-ordination is achieved by substances called gamones released by the spawning animal or by the gametes themselves which induce other individuals in the neighbourhood to spawn. This produces a chain reaction and mass spawning devlops.

In general temperate and northern animals have well defined breeding seasons that correlate with sharp seasonal changes in light, temperature or food. This is true of invertebrates living in the sea as well as those living on land and in freshwater. In the tropics land and freshwater animals are affected by the onset of wet and dry seasons. In tropical seas seasonal upwellings of nutrient rich colder waters may be associated with the onset of reproduction in, for instance, bottom living Sea urchins. The phases of the moon have a powerful influence on the sex lives of many tropical marine invertebrates notably those of Palolo worms *Eunice*, tropical relatives of the ragworms *Platynereis* and a Bermudan luminescent worm *Odontosyllis*. The Palolo bristleworms of the Pacific in the region of Fiji and Samoa, and the Atlantic Palolo in the Dry Tortugas form spectacular swarms some eight to nine days after full moon in November. The adult male and female Palolo worms live on the sea bottom amongst rocks or in the crevices in coral. They are rather like ragworms but have more tentacles on the head and have gills arched over their backs. At sexual maturity the sperm and ova accumulate in the posterior segments of the body so that the swollen back part of the worm looks quite different from the head end. At the spawning

time the gamete laden posterior portions of the worms break off and swim to the surface of the sea where they split, spilling ova or sperm into the sea. The egg masses are netted by the people of Fiji and Samoa who cook and eat them. The hind pieces of the worms are still sensitive to light after they have broken off from the head and this helps to guide them towards the moonlit surface of the sea. The head ends of the worms left at the bottom regenerate the segments they have lost. Swarming is important in ensuring that all the members of the population above a certain age come to sexual maturity at the same time and release their gametes in the same place, ensuring maximum chance of the eggs being fertilized. In the ragworm-like annelid *Platynereis* the complete animals swarm around the time of full moon and experimental work involving subjecting different stages to various time periods of illumination have shown that the time of swarming is set very early on in development by exposure to alternating cycles of photoperiod before sexual maturity has even been reached. Interestingly *Platynereis* does not shed its gametes into the water but practises an unusual kind of internal fertilization. The male wraps tightly around the female and pushes his anus into her mouth. The gut of both partners has been digested by cells called phagocytes so that sperms pass directly from the body cavity of the male into the body cavity of the female, where fertilization occurs. The black spiny Sea urchin *Diadema* which is common on Mediterranean coasts and elsewhere, also spawns at full moon but only in the Red Sea; populations of this urchin

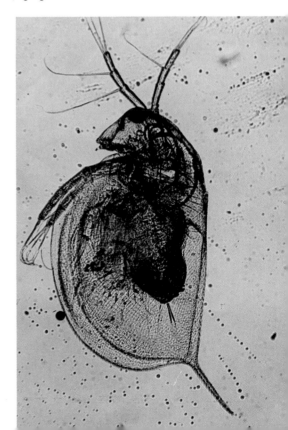

The Water flea *Daphnia* exists in a female form for much of the year and reproduces parthenogenetically without males.

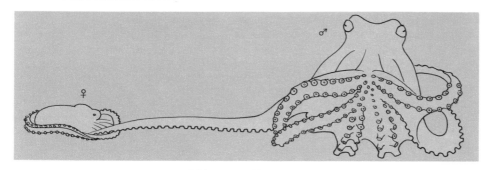

The octopus copulates at arm's length and uses the third (hectocotylized) arm to deposit sperm packets in the mantle cavity of the female.

living elsewhere seem more immune to moonlight and do not have a lunar cycle of spawning.

Sexual Attraction and Bioluminescence. Swarms of odontosyllid worms off Bermuda called the Bermuda fireworm and another species off Vancouver Island are a beautiful sight because the male and female worms are luminescent. Swarming occurs about one hour after sunset on the 2nd to 4th days following a full moon. The female worm swimming in circles near the surface of the sea suddenly produces a rich golden glow on the posterior regions of her body. She starts to throw off eggs and also sheds luminescent material that lights up the surrounding water. The male locates the female by her light and approaches from below, making for the centre of the luminous ring caused by the female swimming round and round as she sheds eggs. The males also light up, producing short puffs of light as they near the surface. The male and female then perform a kind of sinuous 'dance' during which the males shed their sperm into the water. The male fireworms can be tricked into swarming around a flashlight shone into the water after dark. Several squids have light organs or photophores and in some cases it is suspected that these may be used to signal to members of the opposite sex or at least to serve as a species recognition pattern. The Firefly squid of Japan *Watasenia scintillans* has photophores on the tips of its ventral arms, scattered irregularly over its body and on its eyes. The pattern of the photophores in males and females is different, which suggests they could serve to distinguish the sexes. The Jewelled squid *Thaumatolampas diadema* from the depths of the Indian Ocean displays lights of three different colours. The eyes are accentuated by a kind of strip lighting formed from grey and blue photophores, while on the body there are red, blue and white photophores. Unfortunately these squids are difficult to keep in aquaria and we know all too little about their sex lives and the part that luminescence may play in their lives as a whole.

Land living animals that use light flashes as a sexual attractant are the fireflies which are nocturnal beetles. Some fireflies emit a continuous glow, whilst others flash periodically. In forms that flash, mainly tropical fireflies, recognition depends on the critical time interval that elapses between the male flash and the response flash of the female. Closely related species often flash with different rhythms and this time interval serves as an isolating mechanism preventing inter-breeding between closely related species. A recognition system is necessary in this group because the females are often wingless, resembling larvae, so that the male has to be lured to the more or less stationary female. One female Glow worm of the genus *Photuris* seduces males of an unrelated beetle *Photinus* by imitating the flash sequence of the female of *Photinus*. When the hapless beetle gets too near he is eaten by the female deceiver – who presumably makes light work of him! The female alters her signalling slightly to attract males of her own species.

Seductive Scents. Insects also use chemical signals, called pheromones, as sex attractants. These are probably by no means restricted to insects and have been implicated in, for instance, parasitic roundworms that have to locate a mate in the cavernous darkness of the mammalian intestine, but most is known about insect sex attractants. The ability of female moths to attract males over long distances (up to $2\frac{1}{4}$ mi [3·6 km] in the case of the Gypsy moth) gives some indication of the great sensitivity of these sex communication systems. Indeed it has been estimated that only a single molecule of the attractant, named bombykol, produced by the female Silk moth is required to stimulate the male antenna. This means that the males can navigate towards the females from areas of very low concentrations of the attractant.

Diurnal or day flying butterflies and other insects tend to use visual signals for sexual recognition and a fine balance is kept in an evolutionary sense between being visible to a potential mate and being visible to a predator. Scent scales are also used as a perfumed sex attractant in butterflies. Cuttlefish are amongst the most intelligent invertebrates and use colour patterns as behavioural displays of various kinds. The brown and white zebra stripe threat patterns of displaying males are particularly distinguished, but all too little is known about the behavioural repertoire and its significance in these beautiful animals.

Courtship and Copulation. The actual process of coming together of the two sexes and sperm transfer is quite a tricky business. One has only to look at the incredibly complex copulatory organs of, for instance, the flea or even the hardened tube-like penises of some fish parasitic flatworms (Monogenea) to begin to see what might be involved. The copulatory organs are in many cases attachment organs as well as organs of insemination, and hooks, barbs, spines and clasping devices are often incorporated into their design. Some planarians and Sea slugs (for instance *Alderia modesta*) use unorthodox channels for introducing their sperm and simply thrust the penis tube through a conveniently exposed flank. The more conventional male whelk has an unexpected immodestly-sized, unarmed penis several inches in length. Some kinds of marine snails, earthworms, leeches, squids, octopuses and cuttlefish are amongst the animals that have no copulatory organs. Aphallate invertebrates frequently exchange packets of sperms called spermatophores. Male cephalopods produce complex spermatophores and use a modified arm to transfer them into the mantle cavity of the female. In the squid *Loligo* the sperm packets are long cigar-shaped structures about 0·5–0·75 in (12–19 mm) long, capped at one end with a tapering filament. Inside the outer case is an inner sperm containing case that occupies about three-quarters of the length of the outer case. The end of the inner case is tipped with cement and fluid is present between the two cases. As they are produced these spermatophores are stored, sometimes 400 at a time, in a special pocket of the reproductive tract. Copulation is preceded by a courtship ritual that involves the male squid swimming alongside the female stretching its arms and assuming a dark red blush. The male then clasps the head of the female and uses the 4th arm on the left side, which is shortened and has modified suckers for attachment of the spermatophores, to pluck out a bunch of spermatophores which are inserted into the female's mantle cavity. Here they rupture and the inner capsule containing the sperm becomes attached near the oviduct by the cement cap. The sperms themselves do not become activated until egg laying occurs. Squid have special spawning grounds and females lay egg masses onto a communal egg pile. Octopuses keep their females at arm's length during copulation and merely stretch out the modified arm or hectocotylus which has a spoon-shaped tip and a groove for the passage of the spermatophores. This arm is at first used to caress the female and is then

The larva of a Glow worm *Lampyris noctiluca* feeding on a snail. Glow worms are beetles and the female often retains the larval form and has no wings. She thus has to lure the winged males with light signals.

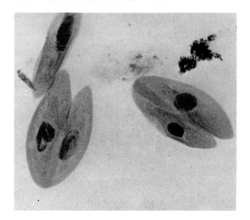

Sexual reproduction in the ciliate protozoan *Paramecium* seen conjugating in pairs. Mating is only possible between compatible mating types.

placed inside the mantle cavity and sperms deposited near the opening of the oviduct. The pelagic Violet snail *Ianthina* which drifts around the seas on a bubble raft it secretes itself has a remarkable way of insemination. Like the cephalopods, Violet snails lack a penis but instead produce large carrier sperms with degenerate nuclei to which other sperms attach. These carrier sperms can swim strongly in sea water and probably convey the 'passenger' sperm directly into the oviducts of the females.

Surprisingly some of the most hazardous mating partners are found amongst the ciliate protozoans. The slipper-shaped ciliate *Paramecium* undergoes a kind of sexual process called conjugation which involves temporary fusion of the contents of the cells of two individuals and exchange of cytoplasm and genetic material. Mating is possible only between compatible mating types. Each species of *Paramecium* has several varieties and within each variety there may be as many as eight mating types. Fortunately paramecia seem in general to be able to pick out suitable partners but some apparently quite suitable mates provide an unsuspected hazard. These partners in *Paramecium aurelia* contain bodies called mu particles in their cytoplasm and can cause death in a sensitive strain of *Paramecium* that receives the mu particles during conjugation. Mu infected individuals are called mate-killers. Other kinds of similar particles occur in different strains of *Paramecium* and there is some evidence that some of them are highly specialized bacteria. Another 'particle' called lambda has been cultured artificially and has been found to synthesize one of the B vitamins, this possibly being of advantage to the host *Paramecium auratum*. The possible advantage of yet another 'particle' called kappa which causes so-called killer paramecia infected with it to release a substance into the water that kills sensitives, as it were, at a distance, is more obscure.

Female crayfish carry their eggs and young around with them on specially adapted limbs. Here baby crayfish cling to the underside of their mother with hooked pincers.

make a good brood chamber for eggs and has developed an extraordinary relationship with these mussels. The female bitterling only ever lays her eggs into the mussel and has a special long, tubular ovipositor to lay the eggs into the inhalent siphon. The mussel does not retract its sensitive siphon when touched by the fish because previously the male bitterling has gently butted this with his nose to accustom the mussel to their presence. Once the eggs have been deposited, the males shed sperm which is sucked into the mussel with the respiratory and feeding currents. The fertilized eggs develop, along with the mussel's own larvae, to young fish which remain in the mussel until they are capable of fending for themselves. The relationship is not entirely one sided, however, because the mussel shoots out its glochidia larvae at bitterlings swimming in the neighbourhood prior to depositing eggs so that the fish become covered with parasites.

The European oyster Ostrea, but strangely not the Portuguese, Japanese or American oysters Crassostrea, brood their larvae in their gills in a similar manner. The oysters are said to be 'white sick' as the eggs first accumulate in a creamy mass, then 'black sick' as pigment is laid down in the developing larvae. The larvae are released after about two weeks and are ciliated veligers (see later) with a small shell. Brooding the young must considerably increase their chances of survival because

the European oyster produces only 1·8 million eggs whilst oviparous oysters that lay eggs directly into the water produce 100 million. The freshwater snails such as Viviparus, Hydrobia and the tropical Spire snail Melania, which is often present in tropical fish tanks, are viviparous and in the River snail Viviparus the eggs are retained in the oviduct of the female, young shelled individuals finally emerging. Some bristleworms called Scale worms that have plate-like extensions of the body fitting together like tiles over their backs also brood their young under the 'scales'. Another annelid worm related to the fanworms, called Serpula, which has a calcareous white tube and is often found encrusting empty shells cast up on to the beach, incubates its young inside its tube.

Many invertebrates that lay eggs (ie are oviparous) protect them in a cocoon and this is typical of land-living and freshwater invertebrates. Planarian flatworms lay eggs together with yolk cells into cocoons measuring 0·04–0·12 in (1–3 mm) in diameter. The cocoons consist of dark brown resistant quinone tanned protein and have a characteristic shape, sometimes being stalked. These features help to identify cocoons produced by different species of planarian. One to twenty eggs may be present in the cocoons which are glued under stones or to the undersides of leaves in the pond or stream. Earthworms, leeches, and spiders are other

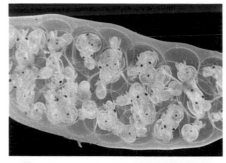

The early stages in the life of a squid. A portion of egg mass showing developing squids (top), and (bottom), a young newly hatched squid.

land living forms that protect their developing offspring from drying up and from attack by bacteria and predators inside a cocoon. The cocoon of earthworms is produced by the glandular clitellar region and slips forwards off the worm collecting eggs and sperm as it does so. Fertilization occurs in the cocoon which when hardened is about the size of a pea but is slightly pointed at each end and is dark brown. Although several eggs are laid into this together with thick albumen secretions, usually only one egg survives. The young worm emerges after 1–5 months. Leeches produce similar kinds of

The remarkable metamorphosis of a Sea star where the tiny starfish separates from the filmy larval body that at one time carried it while it swam in the plankton.

cocoons. The freshwater snail-eating leech *Glossiphonia*, carries several cocoons attached underneath the body and when the young hatch they remain attached to the parent and are carried around for a while. The European bloodsucking medicinal leech *Hirudo*, which in the past was used in blood letting to cure ailments as diverse as a fever or a black eye, lays large spongy cocoons nearly an inch (2–3 cm) long, depositing them in damp places on land. Spiders weave silk cocoons for their eggs and these are sometimes camouflaged. Some spiders abandon their cocoons, others guard them, some even carry them around attached to their spinnerets. Wolf spiders carry the cocoon around and when the young spiderlings hatch they climb onto their mother's back and cling there for a week or so before dispersing. The heathland spider *Achaearanea*, which makes an untidy web that serves as a sprung trap for ants guards its egg cocoons in a domed retreat at the top of the web; when the temperature rises in the heat of the day the solicitous mother carries the cocoons out to air in cooler parts of the web, dutifully returning them as the temperature drops later. When the eggs finally hatch the female carries ants to the retreat for herself and her offspring. Another European spider, *Theridion sisyphium* regurgitates food and allows the young to feed from her mouth, whilst *Coleotes* sacrifices itself to the offspring and after guarding them for a while the emaciated female is eventually devoured by them.

Certain kinds of crustaceans such as shrimps and woodlice also carry their young around with them, although no cocoon is produced. Shrimps develop special egg carrying setae or hairs on the limbs at the breeding season and eggs are cemented to these. Shrimps, lobsters and related animals carrying their eggs around in this way are said to be 'in berry'. The woodlouse, *Oniscus*, has a special broodpouch under the body in which the eggs and young are carried.

Female octopuses are quite attentive parents and hang their fertilized eggs in strings from the roofs of the caves or crevices they occupy, remaining on guard without eating in order to aerate them with jets of water and to carefully clean the eggs of encrusting organisms with their tentacles. A small floating octopus, misleadingly called the Paper nautilus because the female sits in a thin white papery 'shell' uses its shell as a perambulator for the egg strings. An additional function of the 'shell' (which is secreted by the arms and is not a true shell) may be to carry the 1–2 in (2·5–5 cm) reduced

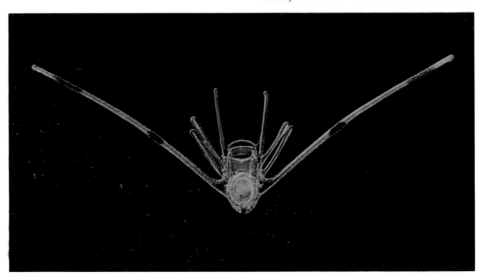

Sea urchins and brittlestars have a delicate long armed larva which floats and feeds in the plankton. The long arms are supported by a calcium skeleton and help the larva to float.

male around. Strangely, some kinds of Sea stars brood their young. Some curve their undersurface and bend the arms down – standing on tip-toe as it were to form a chamber for eggs underneath. Others brood the eggs in the angles between the arms.

Of all the invertebrates parental care is probably best developed in insects, ranging from the situation in solitary wasps where the parent captures and paralyses other insects, stores them in a hole and lays an egg on top forming a kind of pantry for the larva, to social insects like the bee where young are reared and fed by 'nannies' (the workers) and caste is decided by the kind of food the young receive. Some of the more unpleasant kinds of parental provision involve the laying of eggs into the living tissue of other animals. This happens notably in ichneuman flies where the young parasitoids kill by eating out soft bodied larval insects like caterpillars into which they were laid as eggs. Screwworm flies, botflies and even ordinary blowflies also lay eggs into living tissue, usually that of man, the skin of game or domestic stock. Horrible wounds can result when the maggots hatch and eat the living flesh. Where larvae, for instance those of insects, are specific to various kinds of food or food plant, the parent has to select the correct food material on which to lay the eggs. The female Cabbage white butterfly is attracted to the mustard oils in cabbage plants and lays her eggs on the food plant. Other animals that select the appropriate place for development of their young are the Garden snail which protects the eggs against desiccation by laying them in the ground and the Dog whelk which cements its vase-shaped egg capsules to rocks and other objects on the sea shore.

Drifting Larvae. It is difficult to imagine, looking at crawling ragworms and tubeworms, or the armoured hangers-on of the sea shore, the limpets, bivalves, barnacles and Sea stars, that these spend their larval lives as delicate, almost transparent planktonic animals drifting, pulsating and dancing in the illuminated surface waters of the sea. So different are many of the diaphanous larvae from the adults they will become that when they were first observed by naturalists they were named as separate animals. For instance, the larva of the Sea cucumber was once called *Auricularia*; it is now termed the auricularian larva of a Sea cucumber. The reason that the larval stage is so different from the adult is because they are specially adapted to their drifting way of life. Transparency is an advantage because it camouflages them, a weirdly shaped, flattened body pulled out into ciliated frills or lobes increases buoyancy and so does possession of outstretched spidery limbs and antennae and projecting bunches of fine hairs and bristles in crustacean larvae. Other buoyancy devices are sometimes incorporated. Oil droplets are stored or perhaps light salts like ammonium chloride are secreted into special pockets or vacuoles.

Because the larva has become specialized to its own way of life just as the adult is specialized to its more sedentary life, the divergence between the two stages has become so great that most larvae have to undergo a drastic reorganization of their bodies, a process known as metamorphosis, to attain the adult state. Metamorphosis is an abrupt discontinuity in the life history, almost like being reborn as another kind of animal. It occurs of course not only in the lives of marine animals but also during

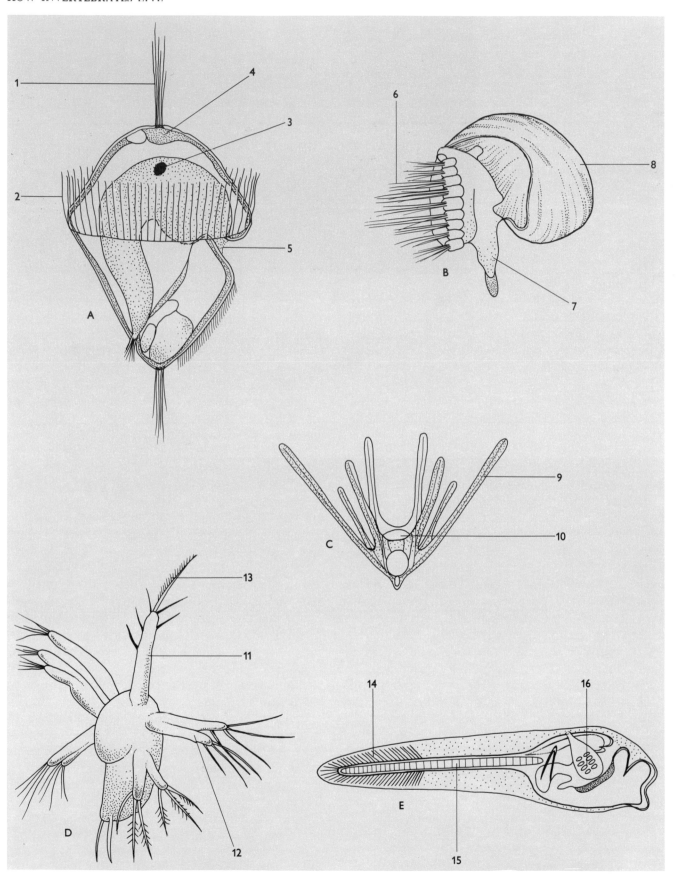

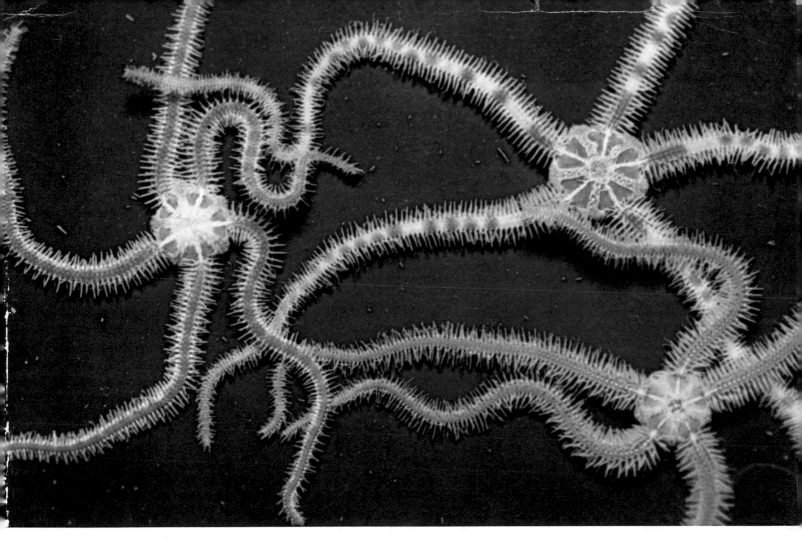

Brittlestars are related to ordinary Sea stars but have much longer thinner arms and a different kind of larva from them.

the development of insects, in the larval transformations of various invertebrate parasites and in the vertebrates, in the development of frogs from tadpoles. The change from one form to another is often so drastic that most of the larval body may be discarded, the adult developing from only a small part of the original larva. This occurs, as will be described, in Sea star metamorphosis.

The reasons that bottom living or sedentary invertebrates go through a drifting (or pelagic) larval stage are mainly so that dispersal of the species can occur and so that the young stages can

◁ **Some of the types of larvae found in the invertebrates. (A) Trochophore of an annelid worm showing the tuft of cilia (1), the girdle of cilia (2), eyespot (3), plate of nervous tissue (4), and the mouth or prototroch (5). (B) Veliger of a mollusc showing the cilia (6), foot (7) and larval shell (8). (C) Auricularium of a Sea urchin showing ciliated bands (9) and the mouth (10). (D) Nauplius larva of a crustacean showing the antennule (11), the biramous limbs (12) and the setae (13). Finally (E), the tadpole larva of a Sea squirt showing the tail (14), notochord (15) and the pharynx with its gill slits (16).**

feed and grow in an environment where food is abundant and where they will not compete with the adult stage. A further function of these fragile larvae is selection of a suitable settlement site for development to the adult stage. The surface of the sea is an ideal feeding ground as these illuminated waters support the minute photosynthesizing algae that are the primary producers on which all other life ultimately depends. These top levels of the sea are a soup of larval forms, some feeding on plants directly, others being carnivores and all themselves being an important intermediate in the food chains of larger animals. The enormous loss in numbers of larvae before development has been successfully completed can be imagined, so the adults have to produce millions of larvae for some to have a chance of surviving predation and locating an appropriate settlement site. It would be imagined that the kind of brooding behaviour practised by the European oyster would be at a premium, but the population dynamics of the situation are always carefully balanced. It would be as lethal to the

107

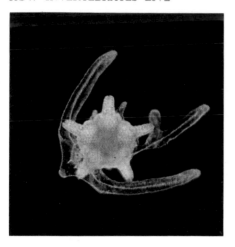

Ophioplutens larva of a brittlestar in a late stage of metamorphosis. The young brittlestar is about to exchange a drifting life at the sea's surface for a more sedentary existence on the sea bed.

convenient position to collect food transported in currents produced by action of the cilia. Very little yolk is incorporated into the egg so that, its food stores rapidly exhausted, the larva has to start filter feeding early in life. As the larva grows, more girdles of cilia are added to support the weight of the trochophore. After feeding for a while in the plankton the larva starts to bud off segments from the hind end towards the head and becomes increasingly wormlike. At this stage the chaetae (bristles) on the last segment of the larva of *Sabellaria*, the Sand mason, become very long and probably help in flotation as well as to deter predators. As the larva feeds, grows and continues to bud off segments, it gradually sinks until, probably in response to encountering the correct surface on the sea bed, it rapidly casts off its ciliated bands and other larval structures and takes up residence as a bottom-living worm.

species to produce too many successful offspring that would survive to exhaust nutrients at a later stage as to produce too few!

Some Invertebrate Larvae. Once the egg of a marine Bristle worm, for instance the ragworm, lugworm or fanworm has been fertilized in the sea it rapidly begins to develop and within 1–2 days, depending on the temperature, has turned into a top-shaped ciliated larva called a trochophore. The trochophore has a long tuft of cilia at the top and a girdle of cilia around the widest part of the body. This larva often has two pigmented eyespots and there is a plate of nervous tissue under the apical tuft of cilia. The mouth is situated under the ciliary band (called the prototroch) girdling the body in a

Molluscs including the Coat of Mail shells (chitons), limpets and some bivalves also have a trochophore stage as their first larva and this shows that the unsegmented molluscs probably arose from primitively segmented worm-like ancestors. The trochophore stage is short lived however and rapidly transforms into another kind of ciliated diaphanous larva called a veliger which occurs only in some molluscs. Veligers have two round head lobes with a fringe of cilia around the edge, a foot region and a gut plus digestive gland region

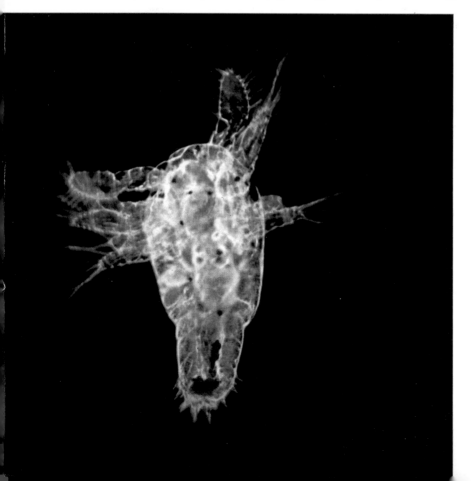

The first larva of marine crustaceans is a pear-shaped nauplius with three pairs of limbs the first of which are antennules. This is the nauplius of the copepod shrimp *Calanus*.

individuals in the male phase to fertilize their eggs. When they have spawned as females they revert to being males again. The number of times this sex change occurs during a year depends on the water temperature and probably on the food available. In cold waters, oysters may change sex only every other year, in British waters they may spawn as males and females in the same year and in the South of France they may change sex several times in the same season. Although individual oysters have only one sex at a time, different members of the population are in different phases so that male and female forms are both present in the same population. Sperms are released into the water by ripe males and are sucked into the mantle cavity of female oysters with the inhalent current to fertilize the eggs.

Certain shrimps also change sex during their life and male copulatory structures are conveniently moulted away without trace as the shrimp assumes its new female identity. Some isopod crustaceans, themselves parasites on the gills of crabs and shrimps, are unusual in that if only one larva infects the host it becomes a female and the second larva to arrive differentiates into a male under the influence of the already established female. Experimental interference with the situation by removing the first established female causes the second smaller individual to become a female. In some arthropods temperature has an effect on sex determination. Free living female Water shrimps of the genus *Gammarus* which are also isopods carry developing eggs in pouches between the legs. If the parents are kept at low temperatures the offspring will develop into males but higher temperatures favour the production of females. Similarly the sex of mosquito larvae can be reversed completely by subjecting them to high temperatures.

Parthenogenesis. Rotifers, Water fleas and aphids are amongst those animals that eschew males and reproduce parthenogenetically by a kind of virgin birth during the summer months. Water fleas are small freshwater crustaceans with a bivalved carapace that swim by means of their long branched antennae. The limbs are used as a filter feeding device and are tucked inside the carapace. In the summer months female Water fleas can lay eggs into a brood pouch situated between the carapace and the body and these develop into small female individuals without fertilization. When environmental conditions become less favourable, perhaps due to drying up of the water, or food shortage, or

Earthworms pairing head end to head end with their tails still tucked inside their burrows for rapid escape if necessary. Earthworms are hermaphrodites, and here they are exchanging sperm.

Slipper limpets *Crepidula fornicata* live attached together in chains and change sex throughout their life. They start life as males and then become female. The individual on the right at the end of the chain is a male, next are intersexed and the individual at the bottom of the pile is an egg-laying female.

a fall in temperature as winter approaches, some of the eggs develop into males. When these males become mature some of the females produce a second type of egg which needs to be fertilized before it will develop. The fertilized eggs often become enclosed in a darkened part of the brood pouch called an ephippium. The ephippial eggs are extremely resistant to unfavourable conditions and can withstand freezing and drying, so are useful to allow the species to overwinter. When the ephippial egg hatches on the return of favourable conditions it will produce a female that reproduces partheno-genetically. In the Arctic the summers are so short that female Water fleas emerging from ephippial eggs may produce ephippial eggs immediately without the intervention of males. These crustaceans have adapted to their environment by eliminating males altogether.

So it can be seen that the 1 to 1 male to female ratio which we tend to think of as normal is by no means the rule in invertebrate animals. The sex of invertebrates is often not fixed and the population dynamics of sex can be very complicated. Even in animals where the sex is fixed throughout life there may not be a 1 to 1 sex ratio. In soil living round-worms (nematodes) where chances of encounter between the sexes are high because they occupy pockets of limited extent, the number of males relative to females is often reduced. This is obviously in the interest of biological economy since when they have performed their appointed task of insemination, males are of no further use to the species and if kept alive only consume food materials that are needed by egg producing females.

Male animals, as a rule, are smaller than females and are usually shorter lived. This again is economical in animals where males do not have a special social role like defending a territory or directly defending the female and young, or competing with other males for females. Many parasitic molluscs and crustaceans have dwarf males which may be permanently attached to the females. The lifespan of animals in general is very much related to reproductive usefulness from the point of view of the species as a whole and many vertebrates as well as invertebrates either age rapidly and die once their reproductive years are over or reproduce only once and then die. The mayfly which mates and lays eggs all in one day and then dies is an example of an invertebrate that does this.

Seasonality and Reproduction. As everyone knows, Spring has a decided effect on sex. This resurgence in temperate climes affects not only land animals but also those in the sea and in freshwater where increasing warmth and longer hours of daylight bring plants and animals alike into breeding condition. Comparatively little is known about the control of sexual activity in invertebrates other than insects but there are indications that in many animals light, warmth and possibly better feeding have an effect on the nervous system causing the production of neurosecretory hormones which then act either directly or indirectly on the gonads, causing them to become mature. It is well known that day

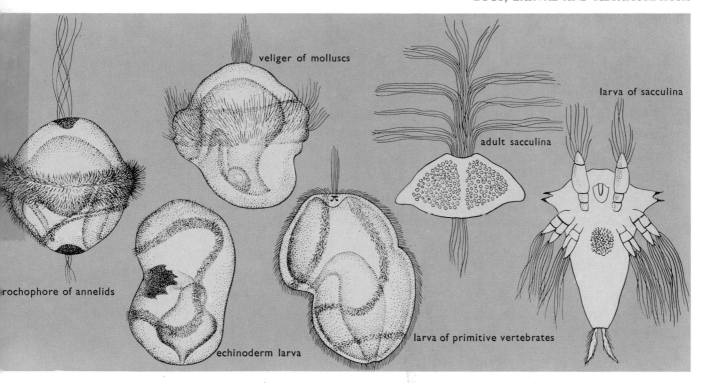

veliger of molluscs

larva of sacculina

adult sacculina

rochophore of annelids

echinoderm larva

larva of primitive vertebrates

The top-shaped trochophore of annelid worm is similar to the first larva of a mollusc indicating that the two groups are related. The echinoderm larva (auricularian of a Sea cucumber) has been involved in theories linking the echinoderms with the vertebrates. *Sacculina*, a parasite on crabs (right) is hardly recognizable as belonging to any group but its nauplius larva gives the game away and indicates that it is a crustacean.

covered by a horny larval shell. By no means all marine molluscs have these dispersive larvae. Most whelks, some winkles and all cephalopods for instance retain the larval stage in the egg or omit it altogether so that the young hatch directly as small copies of the adult. Larval forms are suppressed in freshwater and land living invertebrates as a whole since in running freshwater they would tend to drift downstream and out to sea. On land larvae would be a positive hindrance and would rapidly dry up so most land living invertebrates have a yolky egg protected in egg shells and membranes that develops straight into a replica of the adult. Like all land molluscs, slugs and snails have lost their veliger stage and hatch from a yolky egg. Even amongst those molluscs that have veligers, these may have a very different biology in different animals. Some veligers are produced in enormous numbers and feed in the plankton on diatoms and algae for many months, others feed as larvae but remain afloat only for a week or so. A third kind of larva equipped with masses of yolk and therefore very clumsy in the water, stays only a matter of hours to days in the plankton and is produced in much smaller numbers. Arctic molluscs tend not to

produce a free swimming veliger stage because the cold waters do not support much phytoplankton for them to feed on.

The veliger larvae of all marine gastropods, that is snail and limpet-like forms, undergo a remarkable process called torsion in the larval stage whereby the whole of the top of the body twists round through 180° on the foot. This brings a cavity containing the gills and an olfactory organ that was

The delicate zoea larva of a crab has long spines on its thorax and near the mouth to enable it to float. In the tropics where the density of the warm water is low, pelagic larvae need even longer and more numerous spines if they are to float successfully.

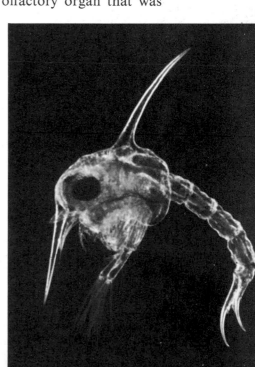

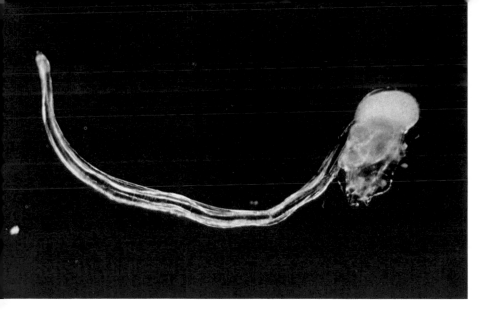

previously at the back end of the animal, over the mouth. The reason for torsion is not known but the most likely view is that of Vera Fretter who suggests that originally torsion was of advantage to adult Sea snails because it brought the gills and olfactory organ to the front of the body so the animal could monitor the environment into which it was moving. Later, she suggests, the timing of torsion was adjusted so that it occurred much earlier and became advantageous to an earlier larval stage allowing it to withdraw the ciliated velum into the now forward facing cavity so that the larva could drop like a stone very quickly in the water and avoid predators.

The Sea stars, Sea urchins and Brittle stars have larvae that are quite different from trochophores and veligers and this shows that echinoderms are not closely related to the annelids and molluscs. Echinoderm larvae have a very frilly appearance due to a convoluted ciliated band that runs around the body being pulled into four basic lobes at the 'corners'. Each group of echinoderms has its own recognizably different larva. The Sea cucumbers, as mentioned, have an 'auricularian' larva so called because the lobes of the ciliated bands form ear-like flaps on each side. The Sea urchins have a larva

shaped like an umbrella upside down but with four or so very long fine arms supported by hard calcium rods pointing upwards, these being edged with a band of cilia. Sea star development is very dramatic. The ciliated larva gradually develops an extremely ruffled veil-like edge and the ciliated band above the mouth fuses into a closed loop. The ciliary extensions are much more flexible than in the Sea urchin larva and are not supported by skeletal rods. Just before metamorphosis three long arms grow out. These have sticky tips which fasten down the larva (now called a brachiolaria on account of the arms or brachia). The young Sea star then develops very rapidly at the back end of part of the larva and separates from it; the filmy larval body is then discarded.

The larvae of crustaceans are more similar to crustacean adults than echinoderm larvae are to their adults. These pellucid creatures have a thin cuticle; jointed limbs and antennae often bearing fine setae are present from an early stage. The first larva of crustaceans is typically a pear shaped nauplius with three pairs of limbs, the first of which are antennules. The nauplius moults several times and is succeeded by various larval types which differ, depending on the group of crustaceans before

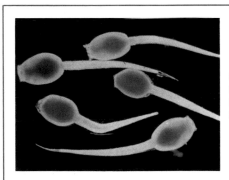

Sea squirts have a remarkable larva that looks like a tadpole (called an ascidian tadpole larva). This larva suggests that we (vertebrates) may be more closely related to sea squirt-like forms than to any other invertebrate group.

Acorn barnacles *Chthapnalus stellatus* can be a nuisance when they foul the hulls of ships or the cooling ducts of power stations. The places they settle in is determined by the behaviour of the larval barnacles.

the adult stage is reached. Crabs, for instance, have a zoea stage which has a jointed abdomen, a spined carapace behind the head and large eyes. The glassy zoea of the Porcelain crab is a remarkable sight for the animal seems to be hanging from a thin horizontal strut. This in fact is a unicorn-like spine many times the length of the body which helps in flotation and probably prevents the larvae being easily eaten by a predator! The stage succeeding this is a megalopa with a square chunky body, reduced

tail and a pair of claws. Freshwater crustaceans tend to have direct development, as is the general rule, and the Fish louse *Argulus* is one of the very few freshwater crustaceans to have a nauplius larva.

Of all these larvae one of the most interesting to the zoologist is that of the Sea squirts. This larva has a round blunt head and sharply demarcated tail and is called an ascidian tadpole because of its tadpole-like shape. The tail is lashed from side to side to propel the larva. Along the tail runs an

Transparent, pelagic nauplius larvae of Acorn barnacles. These larvae will develop into bivalved cypris larvae which determine where the adult barnacles will become established.

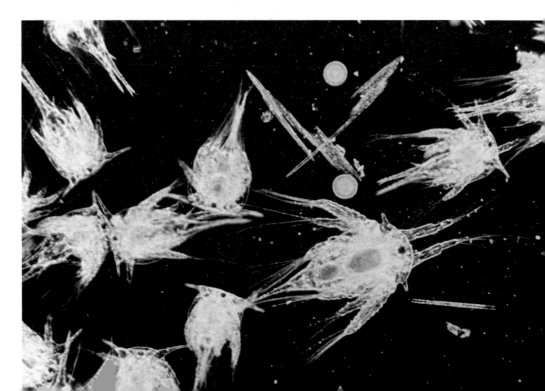

Stages in the settlement of the American oyster *Crassostrea virginica*. **The larvae respond to secretions produced by adult oysters already present: when one settles the other larvae quickly follow. Larvae swimming in the plankton (1–3), beginning of settlement (4), crawling with the foot (5–7), fixation with the left shell valve (8), and settled (9–10).**

elastic strengthening rod, the notochord, to which muscle blocks are attached. Above this runs a nerve tube that connects with a dorsal 'brain' in the head. This is very unusual for an invertebrate since these usually have a solid nerve cord running along the belly rather than along the back. This larva also has slits in the pharynx broadly similar to the gill slits of fish. The similarity of the ascidian tadpole larvae to Chordates (the group that includes the vertebrates) has led to zoologists classifying the Sea squirts as Protochordates. Notice that this similarity is hardly noticeable in the adult, except for the pharyngeal slits which are multiplied and form the meshwork through which Sea squirts strain water during filter feeding. It is true of many larvae in fact that they tend to reveal relationships that are not obvious from studying the adults. When the ascidian tadpole metamorphoses into an adult Sea squirt it cements down the head and casts off the tail so the adult is formed only from the head region. In one group of ascidian tadpoles the larva has become sexually mature, producing a new kind of adult that looks like the tadpole larva rather than a Sea squirt. This is exactly analogous to the way in which the larval stage of certain amphibians has become sexually mature in the external gill stage to produce a new kind of adult, the axolotl. The ascidian tadpole-like adult *Oikopleura* secretes a mucous 'house' and hovers inside lashing its tail to

An early method of oyster culture practised in Japan, the spat settling on bamboo fences and being raked off when full grown. Modern methods (right) of collecting and growing oysters include strings of old oyster shells hung from rafts and also piles of curved tiles stacked on the beach.

The European oyster *Ostea edulis* exposed at low tide has been cultured since Roman times but many oyster beds are now in danger due to overfishing and pollution.

produce water currents. Two fine screens are built into the side of the mucous tent which allow only the finest particles to enter. Inside, small planktonic animals are caught on sticky threads and eaten. The process whereby an adult animal retains the form of its larva is called neoteny.

Settlement. The pelagic larvae of marine invertebrates are able actively to detect those sites which will be most advantageous to them in terms of food and shelter in their future adult life. In some cases they are able to postpone metamorphosis, sometimes for long periods until they find an appropriate site. Settlement is no hit and miss affair therefore but depends on the larvae recognizing certain clues that guide them to a suitable substrate, that is, a suitable part of the sea bed. The economic importance of Acorn barnacles as fouling organisms on the hulls of ships and in the cooling conduits of power stations has led to much interest in the factors that bring about larval settlement. The larva that hatches from the eggs of barnacles is a pear-shaped nauplius which is a dispersive stage feeding in the plankton. After several moults the nauplius changes into a bivalved cypris larva with eyespots and it is this stage that searches for a suitable attachment site. The cypris is not free swimming but tends to crawl over the surface of rocks or other structures. It is sensitive to water currents and weak currents encourage attachment although stronger currents prevent this. Concave surfaces stimulate settling presumably because they offer some protection. The main settlement stimulus, however, is chemical and is caused by contact with the quinone tanned protein forming the cuticles of already settled larvae of adults or even the old bases of dead barnacles. The stimulus is not absolutely specific and the barnacle *Balanus balanoides* is induced to settle by three other species of the same family. Despite this chemical attraction, barnacles tend not to settle too densely on top of one another but space themselves out effectively. The gregarious settlement of barnacles is advantageous to the species because adult barnacles have to be fairly close together in order to copulate. Barnacles are hermaphrodite and in their male role individuals stretch out a long 0·5 in (12 mm) penis, which is longer than usual during the breeding season, and insert this into the shell plates of a neighbour. Also, settlement on a substrate that has already supported a colony of barnacles may be considered as re-exploitation of a tried and tested site. If larvae do not encounter the settlement stimulus within a certain time they may either become incapable of metamorphosis, even when finally presented with the correct stimulus, or may undergo a faulty metamorphosis. Obviously barnacles can settle on uncontaminated surfaces like the newly painted hulls of ships so that settlement stimuli are not absolutely necessary. As soon as one larva makes contact with a previously uncontaminated surface,

113

Adult *Sabellaria alveolata*, the sand-mason, removed from its tube. The larvae of this worm settle where adults are already established.

however, it will encourage others to settle which will rapidly cause a chain reaction as the increasing number of settled barnacles further increases settlement.

The European oyster *Ostrea edulis* has been cultured since Roman times. Cultivation usually involves placing ropes, dead shells or 'parcs' of curved tiles stacked in piles as settlement sites for the larval oysters. After being brooded by the oysters the larvae swim and feed for about two weeks before settling. The first stage of settlement involves the larvae coming into contact with and crawling on the surface of a likely object, anchored by thread-like secretions as it explores. Settlement is again stimulated by the secretions of previously established individuals but spatfall is also influenced by the nutritional state of the larva (which is related to light intensity and temperature) and the length of time it has been swimming. When the larva has settled and metamorphosed into an adult it secretes a powerful cement that attaches the oyster firmly to the substrate.

Gregarious settlement behaviour is becoming established in other invertebrate larvae and occurs in the larvae of the Sandmason *Sabellaria*, a bristleworm that builds extensive sandy reefs composed of millions of tubes, along sandy beaches. The chemical secretions of already settled individuals are also important in causing the trochophore of another bristleworm, *Spirorbis*, to settle out of the plankton. *Spirorbis* is the small coiled fanworm with a chalky white tube often found encrusting the fronds of sea weeds. Not surprisingly the larva of *Spirorbis* has also been found to be induced to settle by extracts of the algae. The presence of bacterial films is important in triggering the settlement of hydroids *Hydractinia* that encrust whelk shells occupied by Hermit crabs and is being implicated as being of importance in influencing

The starfish *Marthasterias glacialis* regenerating four arms from the original disc (body) and with one original arm, a condition commonly called the 'comet' form.

A massive colony of *Sabellaria alveolata*, 1·3 ft (0·4 m) long by 0·8 ft (0·24 m) wide, on a rock exposed at low tide. The thousands of worms in it have taken several years cementing sand grains to build the tubes composing it.

the settlement of other invertebrates such as the bristleworm *Ophelia*.

The larvae of Sea slugs which feed only on particular kinds of prey show a very specific settling response and in fact are only induced to settle by secretions from their prey. The larva of the Sea slug *Tritonia hombergi* settles only on the yellowish white soft coral known as Dead Man's fingers *Alcionium digitatum* and another Sea slug *Adalaria proxima* will settle only on living colonies of the bryozoan Sea mat *Electra pilosa*. Dead colonies exert no attraction on the larvae.

Some marine animals suffer from encrustation by other marine organisms, especially slow moving animals with roughened surfaces which facilitate attachment of the settling larva and favour development of a growth of algal slime. Scallop and oyster shells very often become thickly encrusted with tube living bristleworms and sometimes bear barnacles. Sea urchins and Sea stars are particularly vulnerable to marine fouling and have remarkable stalked pincer-like projections all over the body between the spines which contain sense organs and grab invading organisms, so keeping a clean body surface. There are four main types of pedicellaria, some with three blades and some with two. The round so-called globiferous pedicellariae have poison glands to incapacitate other organisms. The bryozoan Sea mats mentioned above which are themselves encrusting animals found on rocks and sea weeds, have independently evolved cleaning structures analogous to those of echinoderms. Sea mats are colonial animals consisting of hundreds of fused rectangular or spherical skeletal chambers each of which is occupied by a tiny filter feeding individual. Some of the chambers are occupied by truly science fiction-like individuals called avicularia. These resemble a transparent bird's head in shape and have a powerful curved 'beak' which seizes animals and other objects they encounter. In addition to avicularia there are long mobile bristle-like structures on the surface of the colony which, as it were, brush the Sea mat by sweeping unwanted material away.

115

The Art of Successfully Opting Out

One of the first things noticed by Anton van Leeuwenhoek with his newly developed microscope was that some small invertebrate animals collected from amongst the lichens on the roofs of houses could be dried out and would remain in a state of suspended animation until rewetted. This observation can easily be repeated. There is a red coloured rotifer called *Philodina roseola* that in hot summers dries up almost completely in rain gutters and bird baths. A smear of this apparently inert crumbly red powder examined under the microscope will gradually come to life if moistened with water. We now know that this ability to survive unfavourable conditions in a state of suspended animation is shared by a number of lower invertebrates living in moss, soil or lichens, notably nematode roundworms, rotifers or Wheel animals and a group of rather odd microscopic stumpy limbed, segmented animals with claws called tardigrades or Water bears. The phenomenon was named cryptobiosis or 'hidden life' by David Keilin in 1959 because, as we shall see, the apparently inert dried organisms are not dead but merely dormant with their metabolism ticking over very slowly. Cryptobiosis is very well developed in very simple organisms such as bacteria and many single celled protozoans and also occurs in plants. Bacterial cultures dried in a vacuum may easily survive for 14 years or so and this forms a useful way of storing commercially or medically important bacterial strains. Many animals and plants produce special resistant stages early in development which can become dormant for various lengths of time and enable the organism to survive unfavourable environmental conditions. These often serve as dispersal stages. The seeds of some plants such as the Sacred lotus and the fern *Chenopodium* have been said to have germinated after suspended animation for nearly two thousand years! Certain crustaceans such as the Brine shrimp *Artemia salina*, which can live in drying salt lakes with up to five times the salt concentration of sea water, produce special thick walled resistant eggs that show cryptobiosis. These eggs can be stored in a dry state for several years but when placed in warm brine will hatch to release nauplius larvae that will grow into adults. The ability to produce living Brine shrimps, as it were to order, from the dried stored eggs is one of the reasons that these shrimps are so extensively reared in teaching laboratories. Other animals rely on resistant cysts like the Potato root eelworm, a notorious parasitic nematode causing potato disease.

Remarkably, the unprotected larva of a chironomid or biting midge, that lives in rock pools in Nigeria and Uganda that often dry up, also shows cryptobiosis. These midge larvae are, according to Professor Hinton 'the largest and most complex multicellular animals that can be totally dehydrated without ill effects'. The ability to survive desiccation in this way is of obvious advantage to the midge larva. When the exposed rock pools in which they live dry up the larvae also dry up and remain

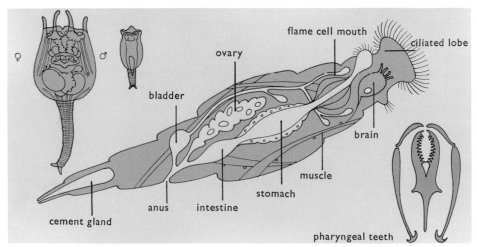

Rotifers, or wheel animals occur mainly in fresh water or on damp soils or moss and are microscopic. They show a remarkable resistance to being dried out completely to a powdery mass. When rewetted they come back to life again. This allows them to live in temporary pools such as rain gutters and bird baths. The males are usually much smaller than the females (top left). The form of the pharyugeal teeth (bottom right) serve to distinguish between the different species.

116

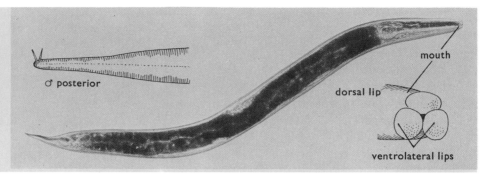

Some free living round worms can tolerate being completely dried out for as long as 39 years. This is an adaptation to surviving severe droughts. When moistened again the worms can absorb water through the cuticle and start to revive.

♂ posterior

mouth

dorsal lip

ventrolateral lips

dormant in the mud until it rains. When the pools fill up again the inert larvae absorb water through their skin until they become fully hydrated and active once more. These pools also contain round-worms that show cryptobiosis. Most living organisms probably survive adverse conditions in the form of special encased spores, eggs, cysts or pupae so it is much more unusual to find apparently unprotected larvae like the midge larvae mentioned above or adults, as in rotifers, nematodes and tardigrades, being able to tolerate complete drying out. The length of time that these animals can survive in a cryptobiotic state is not known. Some adult nematodes have been revived after being kept in a dried condition for 39 years. Moss that had been kept in a museum for 120 years was found to contain a number of rotifers and tardigrades. When these animals were moistened a few revived but died almost immediately, so there is probably a limit to the time that these animals can be dried out for. Rather like Rip van Winkle who awoke to find himself an old man, dormant organisms do not escape ageing altogether, but only slow the process down and so cannot remain alive indefinitely. When

the dormant animal is 'awakened' again by a 'christening' of water ageing recommences at the normal rate. This illustrates the difference between calendar age and biological age. An animal that is one day old biologically (in terms of development) may have a calendar age of one year and a day if it has been in a state of cryptobiosis for one year. It has been estimated that a tardigrade that would normally have a life span of only one year may extend its life expectancy to the ripe old age of 60 by 'taking it easy' and entering into cryptobiotic states at intervals. Biologically, cryptobiosis is useful only if the animal survives to reproduce.

One of the most interesting things about cryptobiosis is that it not only confers resistance to drying, it also makes tardigrades, rotifers and nematodes incredibly resistant to extremes of temperature, to ionizing radiation and to the reduced pressure of a vacuum. Dried rotifers and tardigrades will survive when subjected to temperatures as low as −430°F (−270°C), close to absolute zero produced by cooling in liquid helium. Conversely dried Moss animals also survived being heated to 330°F (151°C) for 15 minutes. The resistance of dormant tardi-

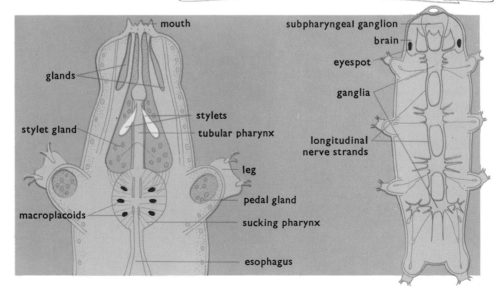

Tardigrades, commonly called Water bears or Moss bears are minute animals with four pairs of stumpy, clawed legs. Most tardigrades live in moss and when the moss dries out in summer the tardigrades are able to themselves dry out and shrivel and to suspend active life. It is a remarkable sight to see them swell up and start to move around when water is added again. The anatomy of the head of a tardigrade (left), and (right), a diagram of the whole body and nervous system.

mouth

glands

stylet gland

macroplacoids

stylets

tubular pharynx

leg

pedal gland

sucking pharynx

esophagus

subpharyngeal ganglion

brain

eyespot

ganglia

longitudinal nerve strands

The female brine shrimp, viewed from the side, lives in salt lakes and produces eggs that can withstand long periods of desiccation. Once the eggs are completely dry they are very resistant to extremes of temperature and can be stored in a jar on a laboratory shelf for several years.

grades to ionizing radiation is equally staggering. It has been found that the enormous dose of 570,000 roentgens from an X ray source is needed to kill 50 per cent of the tardigrades in 24 hours. A human population would show the same number of fatalities if subjected to a dose of 500 roentgens. The resistance of cryptobiosing tardigrades to high vacuum, which is presumably of little biological use to them, has been turned to practical advantage. A dormant tardigrade has been placed inside a scanning electron microscope and subjected both to high vacuum and electron bombardment as the electron probe scanned across. After being viewed for 1 hour it was removed from the microscope and moistened; it revived, but briefly, before dying. Similarly the dried animals are very resistant to mechanical damage. Dry larvae of the midge *Polypedilum* could be cut into several pieces, all of which would survive in a dry state for several years. When wetted again the pieces revived but died shortly afterwards. Extensive damage of this kind to fully hydrated, active, larvae would prove fatal very rapidly.

To enter into a state of cryptobiosis a tardigrade loses 82 per cent of its body water, reducing this to a mere 3 per cent of its total weight. The speed at which it dries out seems to be of great importance in preserving the structure of the complex and vulnerable molecules in the cells. Moss animals control their rate of drying by contracting the body either into a

barrel shape as in rotifers and tardigrades, or by tightly coiling the body like a spring as in nematodes. This not only reduces the surface area over which water can be lost but also covers especially permeable regions of the body wall. Animals that have been anaesthetized and are unable to contract lose water about 1,000 times faster than contracted animals and in fact do not survive the desiccation. The resistance of dormant rotifers and tardigrades to very low temperatures must be more than partly due to the fact that the water content of the cells is so low that damaging ice crystals do not form. The sperm of farm animals is stored in sperm banks for artificial insemination at low temperatures, using much the same principle, except that water is not completely removed from the cells but is substituted by anti-freeze-like substances such as glycerol or dimethyl sulphoxide, prior to cooling. Resistance of dried animals to heat and radiation may be due to the fact that the molecules of the organism adopt an extremely resistant configuration on drying. They may also be protected by some kind of organic molecule that substitutes for water, but this is not yet known. Some of the enzymes must still be in active states since animals in cryptobiotic states still consume minute amounts of oxygen and are presumably respiring.

Land slugs conserve water by being most active at night or after rain. They become dormant in very hot weather as well as in winter.

The ability to survive extreme environmental temperature changes in a dried state is of obvious importance in allowing organisms to live through difficult seasonal changes such as freezing in Arctic tundra or excessive heating in dry tropical regions. It is also important to animals such as nematodes and rotifers that become very light in their dried state and can be dispersed by wind currents.

So far we have mentioned only the ability of

organisms to become inactive due to ordinary drying caused by evaporation. Drying can however also be produced by strong salt or other solutions which draw water from an organism by osmosis. Cryptobiosis in Brine shrimp eggs may be produced by the progressive drying of salty pools. As the water in these pools becomes more concentrated, it exerts a stronger and stronger drying effect on the eggs. Another and rather different physical condition that animals may show a certain amount of resistance to is lack of oxygen. Nematodes deprived of oxygen at first switch their metabolism to anaerobic respiration and after a few days become quiescent. Some species of nematode, but by no means all, can survive the conditions of low oxygen in an inactive state for as long as 100 days and retain food stores that can be utilized as soon as they are returned to aerated water.

Aestivation and Hibernation. Cryptobiosis is rather a special way of resisting adverse conditions and few adult invertebrates are able to survive in a state of suspended animation by reducing the amount of body water quite so drastically. Many invertebrates, particularly those living on land in

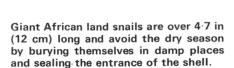

Giant African land snails are over 4·7 in (12 cm) long and avoid the dry season by burying themselves in damp places and sealing the entrance of the shell.

warm climates or in the intertidal zone of the sea shore do however show either long term or short term resistance to drying involving a period of inactivity. Shore living snails, limpets and barnacles, for example, close their shells when exposed at low tide and become comparatively inactive. These activity rhythms will sometimes continue even when the animals are moved into the laboratory and will continue to be synchronized with local tides. Land living invertebrates may go into a dormant state called aestivation in summer or the dry season in order to conserve water. Aestivation occurs, for example, in earthworms and in many kinds of land snail and involves behavioural as well as physiological changes. The surface of earthworms although invested in a latticework of collagen, the cuticle, is permeable to water. This makes the earthworm very sensitive to dry conditions and to avoid dehydration when the soil humidity drops certain species of earthworm can aestivate. There are about 13 species of earthworm in Britain and two of these, *Allolobophora longa* and *Eisenia rosea*, are particularly sensitive to changes in soil water and undergo aestivation. Slow drying out of the soil and lack of food cause the animals to become colourless and to curl up into a ball in an air pocket in the soil. Mucous secretions are produced which dry to coat the lining of the burrow. The worms then become dormant and cannot be wakened for two months or more. The worms emerge from this state much lighter in weight having metabolized food stores and presumably lost some water in their resting state; up to 75 per cent of the body water can be lost without causing death. The dormancy is probably controlled by neurosecretion of hormones from regulatory centres in the brain that in turn receive information from humidity and other kinds of receptor in the earthworm's skin. Other cyclic factors are probably involved in controlling aestivation but these are not yet fully understood. Partly desiccated worms are able to absorb water from moist soil across the cuticle so that they can rapidly regain water lost in dry conditions. Slugs and snails are also able to imbibe water through their skin after drying.

In hot Mediterranean countries aestivating land snails are frequently seen in summer clustering thickly onto bleached grass or on the thin twigs of dusty bushes. With rain these snails will become active again and creep out in swarms. Many of these land snails conserve water by secreting a mucous film or epiphragm over the shell opening which dries and becomes impermeable. Some snails can enter a prolonged state of suspended animation when the humidity drops to an unfavourable level and may burrow into soil where it is cooler to aestivate. Their heartbeat and rate of respiration become slower as they become inactive. A desert living species of snail, *Helix desertorum*, kept in the British Museum is said to have awoken after 4 years when taken into damp air. It is difficult to see how any animal could fail to appreciate the British climate for so long.

Another kind of snail, *Oxystyla capax*, is said to have aestivated for 23 years. The shell of snails acts as a kind of microhabitat taking the brunt of extreme conditions and protecting the animal within. Slugs have no shell to prevent water loss so have to be able to tolerate greater fluctuations in body water. Slugs may lose up to 16 per cent of their body water in an hour by evaporation and in production of that mucus they leave behind a shining trail. The loss is made good by eating and through the skin. Being vegetarians, slugs tend to feed on juicy plant material. Slugs, snails and earthworms also show daily activity rhythms which help them to conserve water. In general all these animals are most active only at night or after rain. It has been suggested that rain drumming on the shell of the snail causes them to become active but whether this is true or not is not known. During the day snails and slugs are fairly inactive and remain under stones or attached to vegetation. Earthworms tend to remain in the soil during the day. These diurnal and seasonal activity phases may be set by internal clocks governed by appreciation of changes in light intensity, as has been suggested for insects like the cockroach, although aestivation can be initiated experimentally by drying snails. Snails in temperate climates hibernate as well as aestivate and *Helix aspersa* the Garden snail may only be active for six months of the year in Britain. In the Garden snail preparation for hibernation may begin as early as September. The snails congregate amongst the roots of shrubs in or at the bases of old walls or bury themselves in the soil. Snails apparently have well developed homing instincts and return to the same communal roosts. When snails hibernate in winter they increase the concentration of their blood proteins. This lowers the freezing point of the blood and acts as a kind of antifreeze that stops the snail icing up even when inactive. Adult snails may not stir until April, but young snails may sometimes be tempted out earlier on a warm day.

Snails often hibernate under bricks or in old walls. They seal the entrance to the shell with mucus and glue the shell to the brick with the same material. Snails use a kind of antifreeze to prevent their blood freezing during hibernation.

Some snails are hosts for important fluke parasites of man and his domestic animals such as *Schistosoma*, the human Blood fluke and *Fasciola hepatica* the 'killer' fluke of sheep. The biology of the respective snail hosts has considerable influence on the distribution and incidence of the diseases they transmit. Flukes normally undergo a vast larval multiplication in their snail hosts, hosts to which incidentally they are highly specific. However, when the infected snail becomes dormant because either hibernating or aestivating, multiplication of larval flukes ceases as does transmission of larvae from the snail. This means that some fluke diseases are transmitted only at certain times of year when the snails are active. Furthermore, adverse conditions may reduce numbers in a snail population at certain times of year and this factor too will be reflected in the incidence of the disease transmitted. Liver fluke in Britain is transmitted by a Mud snail *Limnaea truncatula* which needs wet conditions and temperatures over 50°F (10°C) in order to be able to reproduce. Rainfall also determines the survival of parasite cysts on the pasture and a forecasting service for fluke outbreaks is now run based on past and predicted rainfall.

Diapause. Whilst aestivation and hibernation apply to the adult animal, diapause is a state of arrested development undergone by a larval stage. The cryptobiosis of seeds, spores and animal cysts and eggs such as the Brine shrimp eggs is an extreme form of diapause. Other invertebrates have resting stages that, though resistant are not completely dried out. Most work has been done on the diapausing eggs and pupae of insects and it is known that the timing of the resting stage is determined long before weather conditions become adverse by changes in day length acting on either adults (if a diapausing egg is produced) or larvae (if the pupa is the resting stage). These changes in light intensity may be recorded not through the insects' eyes but directly onto brain cells lying under the semi-transparent cuticle. We also know that particular factors are required to 'awaken' the diapausing eggs or larvae. Simply raising the temperature is not enough. Some overwintering eggs and pupae (like some plant seeds) have to be subjected to a period of low temperature before they will develop further. Insects are not the only invertebrates to undergo diapause. Many kinds of invertebrates have overwintering eggs, soil living rhabditid and parasitic nematodes have a resistant third stage larva and some fluke parasites have eggs or metacercarial cysts that undergo a kind of diapause in the outside world until conditions become favourable for transmission. It is interesting that the eggs of a sheep parasite, a nematode called *Nematodirus battus*, will not hatch until they have been subjected to temperatures around freezing point. This is probably a device to ensure that all the parasite larvae hatch at once in spring when vulnerable lambs are starting to graze. In all cases diapause is not just a way of surviving adverse conditions, it is a way of adjusting the timing of the whole life history of an animal so that it can synchronize with optimum conditions for growth and reproduction. As has already been mentioned, the ability of parasites to gear their life cycles to those of the host is particularly important and subtle.

Colonial Life – Fused Animals

This chapter deals with a special kind of colonial life that is much more intimate than the mere clustering together of individuals as in, for instance, barnacles, mussels or in the social insects. Several kinds of invertebrates form colonies consisting of the fused bodies of many individuals forming an organism by a kind of multiple Siamese twin arrangement. This strange state of affairs only really occurs in fairly simple invertebrates that reproduce asexually by budding. The parent buds off a small copy of itself which grows to full size but remains in contact with the parent and then itself buds so that a large mass of attached individuals is rapidly formed. Unlike the colonies of separate individuals established say in mussels or barnacles, colonies of fused individuals are formed by division of a single individual rather than by the coming together of many larval individuals. This habit of budding off new individuals asexually, that remain in contact, is rather similar to the kind of growth shown by plants and indeed some of the cnidarians in particular produce beautiful branching plant-like forms of extreme delicacy. Intimate colonies of this kind occur in protozoans, sponges, cnidarians, Sea mats and Sea squirts and their allies. The compound animals formed are often very complicated since there is a tendency for members of the colony to become specialized to carry out particular functions. This may mean some individuals are concerned only with trapping prey and cannot feed directly. Other members of the colony may be concerned specifically with locomotion or reproduction. All these individuals may be dependent on certain feeding individuals which do take in food, digest it and pass on soluble food to other members of the colony through the tissue connections that join them all together. Because the individuals have become specialized to carry out particular functions they often look very different from one another. The colony is then said to be polymorphic. Specialization involves members of the colony sacrificing their own independence to the smooth working of the whole colony. The way that this specialization is achieved is not fully understood any more than

A Beadlet anemone *Actinia equina* with newly budded off baby anemones. In this case the buds do not remain in contact to form a colony.

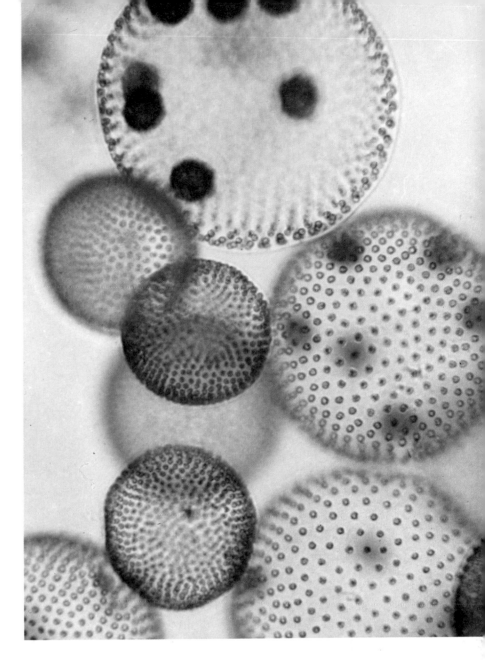

Spherical colonies of *Volvox* may contain up to 20,000 single-celled flagellates embedded in a jelly sphere. The individuals are joined by cytoplasmic threads and the activities of the colony are coordinated. One pole always moves first. Dark spheres inside are daughter colonies.

embryologists understand how a single fertilized egg of any animal divides to produce many different kinds of specialized cells adapted to perform separate functions. It seems likely however that the position in which the individual finds itself within the colony is important in determining what it will become and that a system of chemical gradients keeps different individuals in communication. In both cases the genetic make-up of the daughter individuals (in the case of the colonial invertebrate) or daughter cells (in the case of the embryo) is identical, so different parts of the common genetic programme must be being expressed in the differentiated individuals. A collection of individuals all with the same genetic make-up is called a clone, so each compound animal consists of a clone of fused individuals. These strange compound animals often

have a definite and regular form and only grow to a certain size. They then usually reproduce sexually by producing eggs and/or sperm which fuse with eggs or sperm produced by a separate colony. Sexual reproduction leads to individuals that are genetically different from the parent which will divide to produce whole colonies with a different genetic make-up. This difference in make-up can lead to problems when separate colonies grow towards one another and start to overlap, for instance if the separate colonies are clustered very closely on a rock. Because the overlapping colonies are likely to be different genetically they will not be compatible, are unable to fuse and may compete with one another for space and food. This incompatibility is of the same kind that occurs during tissue transplants when tissue of a different make-up

123

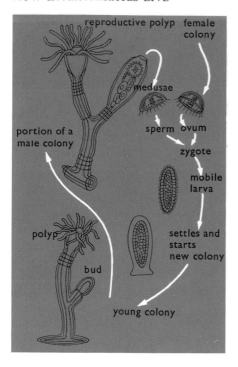

Life cycle of the marine colonial hydroid *Obelia* showing that a sexual phase alternates with asexual reproduction by budding. The branching colony of polyps buds off small jellyfish (medusae) which produce eggs or sperms. These fuse and give rise to a larva which settles to form a new colony by budding.

from that of the recipient may be rejected. Some experimental work has been done on clonal incompatibility in forms like the compound Sea squirt *Botryllus* which forms thin jelly-like colonies encrusting rocks or sea weeds. It has been found that there are about 20 different compatibility types of this animal in any population and only identical types are compatible and can grow together if they meet accidentally.

Most of these colonial animals are sedentary and attached to rocks or some solid support but a few float freely, for instance the Portuguese Man-O'-War, a hydroid colony, and some salps which are barrel-shaped tunicates related to the Sea squirts.

Colonial Protozoans. Some of the best known colonial protozoans occur in the plant-like members

Ostrich-plume polyps form a feather-like colony 3 in (7.5 cm) high. The polyps are borne on the side branches and are protected by a transparent chitinous skeleton into which they can withdraw.

of this group that have chlorophyll and can make their own food by photosynthesis. *Pleodorina* and *Volvox* form spherical colonies. *Pleodorina* consists of 128 individuals arranged at the edge of a ball of jelly and here the individuals are not joined by a cytoplasm. *Volvox* colonies are much larger and according to the species may contain several hundred to as many as 20,000 individuals which are connected by cytoplasmic threads. Colonies of *Volvox* are visible with the naked eye as small green balls rotating slowly at the surface of ponds and other bodies of water. Each individual bears 2 whip-like flagella which beat to propel the colony along. The beat of the flagella is synchronized and there is a definite anterior pole that always moves first and may have longer flagella than on other parts of the colony. These may be sensory and are perhaps used to test the suitability of the water into which the organism is moving. In *Pleodorina* the 'anterior' individuals have larger eyespots than other members of the colony so may show a differential sensitivity to light. Like those of *Volvox* the anterior members of the colony are unable to reproduce so there is a definite division of labour within the colony, separate individuals taking on special roles. In *Volvox* the colony reproduces by budding off special individuals from its wall which come to lie in the jelly filled interior. Here they divide and multiply to form spherical colonies within the parent. When they are first formed these hollow daughter colonies have the flagella inside but as they mature they turn inside out through a small pore. They remain imprisoned within the parent until this dies and meanwhile the daughter colonies may themselves have produced daughter colonies inside. Although colony formation is caused by asexual reproduction, *Volvox* can also reproduce sexually. Different species may either be hermaphrodite or bisexual. The gametes are produced by specialized cells in the wall of the colony and the eggs and sperm will fuse to form a resting zygote with a spiny wall. This is a protective stage which can remain dormant for some time. When it starts to grow and divide it gives rise to a hollow spherical colony which again eventually turns inside out. The advantages of forming this kind of colony involving sometimes thousands of individuals are not altogether clear but probably the main advantage is that the organisms unite their swimming capacities to produce a colonial individual that is much more active. This means the plant-like colonies can disperse themselves effectively in the

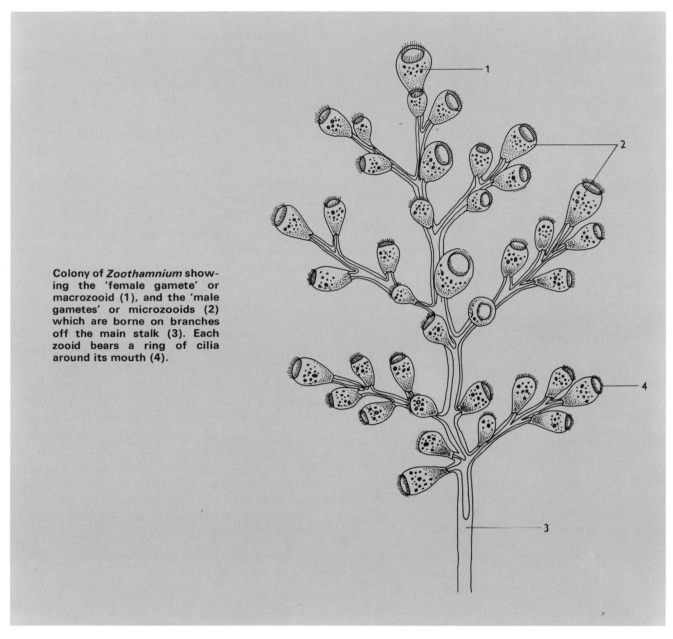

Colony of *Zoothamnium* showing the 'female gamete' or macrozooid (1), and the 'male gametes' or microzooids (2) which are borne on branches off the main stalk (3). Each zooid bears a ring of cilia around its mouth (4).

water and can seek out the optimum conditions of illumination for food manufacture by photosynthesis.

Stalked ciliate protozoans also produce remarkably beautiful colonial forms, this time with a branching shape. Ciliates cannot feed by photosynthesis and catch their own food. Colonies of *Carchesium*, *Epistylis* and *Zoothamnium* consist of a number of round to bell-shaped zooids each with a circlet of cilia around the mouth united by stalked contractile processes. When the colony is disturbed it can contract by throwing the main stalk into a spiral rather like a coiled spring and by withdrawing the circlet of cilia around the mouth that normally

create vortices sucking small food particles into each organism. The way the zooids are arranged on the branching stalks is very precisely determined so that each colony has a particular structure. *Zoothamnium* shows quite a high degree of specialization in its member individuals. It bears at the tip of the colony, like the fairy on the Christmas tree, a specially large zooid which can become the 'female gamete' (macrogamete) in sexual reproduction. The ends of the lateral branches further down bear specialized zooids that can form 'male gametes' (microgametes). New colonies are not only formed by sexual reproduction. Some of the ordinary zooids may suddenly develop an extra posterior

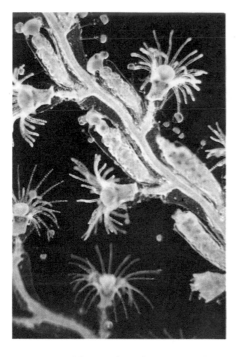

Obelia colony showing how the polyps are connected by a common stem for food-sharing. The vase shaped structures contain developing medusae.

band of cilia and swim off to found new colonies. The advantage of these ciliate protozoan colonies is probably that they allow food sharing and minimize wasteful competition by regular spacing of the individuals.

Sponges may form a special case of colony formation. Some zoologists consider that the whole body of a sponge may be a kind of colony rather than an integrated organism. This idea is based mainly on the fact that no definite nervous system has been found in sponges, although some elongated cells that may have the function of conducting information are present. Apart from this, some sponges may form colonies by budding. The colonial forms are identified because they have several exhalent pores or oscula. The Breadcrumb sponge, *Halichondria* which occurs in shades of green, yellow and orange and encrusts rocks is an example and many large round exhalent pores are borne on volcano-like projections.

Attached Hydroid and Coral Colonies. The colonial hydrozoans like their atypical, solitary freshwater relation, *Hydra*, are delicate organisms. They are usually semi-transparent, often colourless and form branching colonies of two main kinds. Some form upright tree-like colonies with polyps being borne at the ends of the branches, the whole colony being anchored by a rootlet system of hydrorhiza whilst others form flat encrusting colonies with polyps given directly off the rootlet system. *Hydrichthys*, which occurs on fish skin and is one of the few parasitic cnidarians, is an example of an encrusting colony. Many of the tree-like colonies have a thin semi-transparent chitin skeleton around the stem and branches and forming cups into which the polyps can retract. This occurs in *Obelia* a delicate branched colony about 3 in (8 cm) tall found on sea weeds. The body of each polyp is vase or flask-shaped and has a central mouth surrounded by a ring of tentacles. These are carnivorous zooids which catch prey by means of the

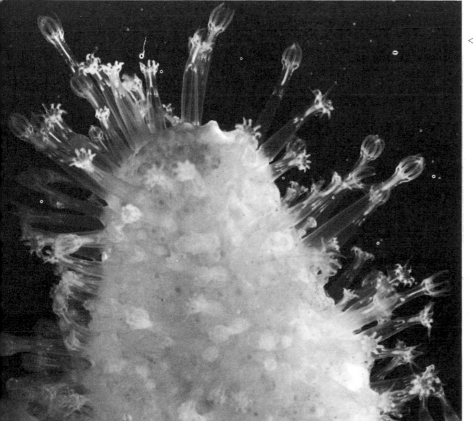

◁ Portion of a colony of Dead Man's fingers *Alcyonium digitatum* which is a soft coral, showing the colonial polyps expanded.

A colony of identical marine coral polyps which at ▷ their bases each secrete a hard cup of calcium carbonate, the coral skeleton.

stinging cells in their tentacles. Food is shared by means of the hollow canal systems that join the polyps. The nervous system probably extends throughout the whole colony because they react as a single individual to stimuli. In some hydrozoan colonies particular zooids are specialized for reproductive purposes and in *Obelia* the reproductive forms are enclosed in a vase-shaped chitinous case and bud off small jellyfish that form one of the dispersive stages. These sedentary hydroid colonies are usually very small but larger colonial forms are found in the Anthozoa which is the group containing the true or Stony corals, the Soft corals, the Sea fans or gorgons and the Sea pens. Most of the Stony corals are colonial and the polyps are connected by a canal system that runs above the hard chalky skeleton they secrete, so that the fleshy part of the coral overlies its massive skeleton. Each polyp sits in a calcareous cup that surrounds its base and has internal partitions that support the

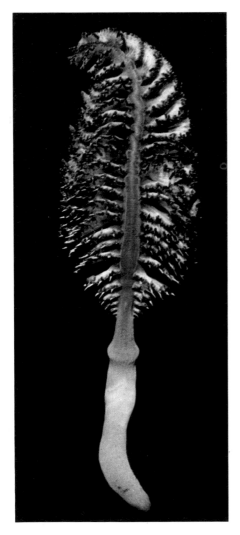

A Mediterranean species of Sea pen gets its name from its quill-like shape. This is in fact a colony of stinging animals (cnidarians) in which the individuals are so coordinated that the colony behaves as a single individual and can, for instance, burrow into sand for protection. The individual polyps are arranged on the feathery branches.

A Sea fan is a horny coral and the orange skeleton is ▷ secreted by thousands of polyps which make up the living part of the Sea fan.

living partitions (mesenteries) dividing up the inside of the polyp's 'gut' cavity. The polyps of Stony corals are all very similar to one another and do not show polymorphism. Dead Man's Fingers (*Alcyonium digitatum*) is a soft coral whose skeleton is often found washed up on British beaches. It has a soft, lobed, orange-yellow body covered with polyps which are partially retracted into the soft matrix. The polyps are joined by canals and share food materials. The soft body of the colony is formed from the swollen basal regions of the polyps which become very fleshy and impregnated with calcium spicules that help support it. The Horny corals include the Sea fans that are often gorgeously coloured in purple, red or orange. They are mainly tropical corals with the exception of Precious coral which is found in the Mediterranean. The Sea fans, as their name suggests, have a flat, branching fan-like shape which is often held at right angles to the main water currents so that the polyps can utilize these currents to trap food. These marine fans are anchored by means of a holdfast and the polyps and their fleshy connecting tissue are borne on the outside of the branches which are supported by an internal skeleton of horny material. One cold water Sea fan is *Eunicella* which is a whitish pink colour and occurs in the English channel. Related to the Sea fans are the Sea pens which are soft bodied despite the presence of spicules in the body wall. They are remarkable colonial animals and consist of a number of polyps arranged on branches coming from a central axis so that they look like a quill pen. The central rachis of the colony ends in a muscular bulb which does not bear polyps. The bulb is used for burrowing and when disturbed the whole colony can fold up and shoot downwards into the sand! This digging ability is brought about by changes in water pressure inside the animal and there is a series of canals running down the rachis into the bulb. During burrowing the bulb is alternately expanded and contracted as water is forced into it and then squeezed out again by muscles in the wall. The overall pressure is controlled by special polyps situated between the feather-like branches on the rachis called siphonozooids. These lack tentacles and cannot feed but can create water currents by means of the long flagella they contain. The polyps on the main branches of the Sea pen are autozooids which can feed and supply food to the rest of the

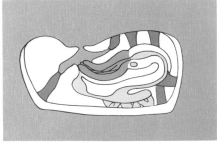

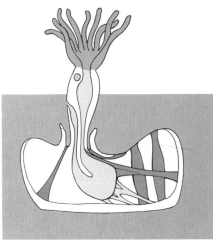

Diagram of a single moss animal bryozoan. The zooid is withdrawn into the box-like home (above). (Below), the zooid is extended with the feeding tentacles (red) visible. Inside the body the U-shaped gut (grey) and nerve ganglia (brown) can be seen.

pneumatophore or float which suspends the colonies at the top of the sea. The Portuguese man-o'-war is really a subtropical form but often strays into the North Atlantic Drift and is occasionally found washed up on European beaches. It looks a little like a jellyfish but in fact is not very closely related to them. Jellyfish are single individuals but the Portuguese man-o'-war consists of many polyps all hanging from a gas-filled float. The float is bluish purple to pink in colour and is about a foot in length. It bears a distinctive ridged crest which acts as a kind of sail and when freshly secreted contains nitrogen, oxygen and, surprisingly, carbon monoxide, which is usually a highly poisonous gas. In rough weather the animal can submerge by letting the gases out of the float through a pore which is usually kept sealed by special muscles. A gas gland at the bottom of the float refills the float for re-surfacing. It is said that the float can be flopped over from side to side into the water in order to keep moist. Under the float are many polyps of three kinds. One kind is concerned with catching food but cannot itself feed. Each of these has a single tentacle which is deep blue and bears stinging cells that can produce a very painful reaction in man. The effect is said to resemble having red hot pins stuck into the skin. The tentacles are enormously long and may dangle 40 ft (12 m) into the water below the float, to provide a very efficient set of fishing lines for the colony. The feeding polyps also have a single tentacle bearing clusters of stinging cells but also have a mouth. The reproductive polyps are lavender

colony. Sea pens are luminescent and some flash irritably when disturbed.

Floating Colonies - Portuguese man-o'-war. Perhaps the strangest colonial animals of all are the siphonophores such as the Portuguese man-o'-war *Physalia* and Jack Sail-by-the-Wind *Velella*. These are hydrozoans related to delicate colonies like *Obelia* but are much larger and have a special

The Portuguese man-o'-war ▷ resembles a jelly fish but is actually a colony of specialized polyps which hang from beneath the balloon-like float. The thick curled tentacles have a deadly sting and belong to polyps specialized to trap food, other polyps actually eat the food and a third kind of polyp neither catches or eats food but is responsible for reproduction only.

◁ This Sea mat colony is commonly seen on European shores growing on rocks or on seaweeds.

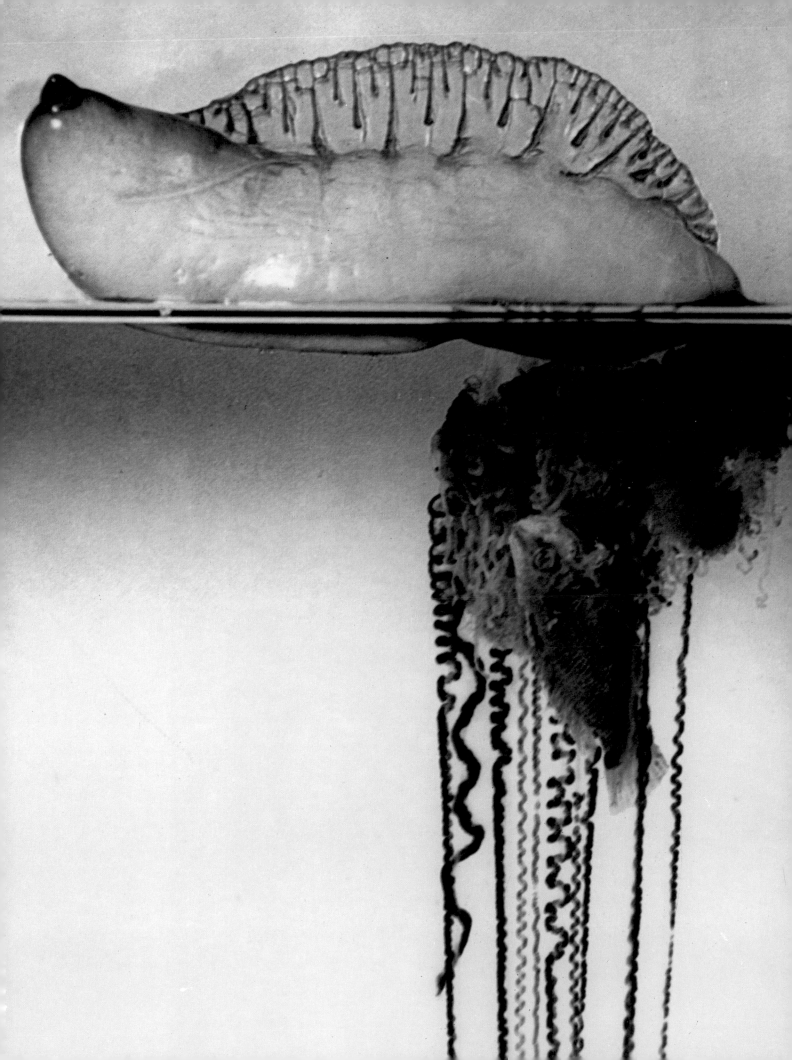

coloured and produce dispersal stages. The tissues of all these polyps are in communication with one another so that the colony behaves like an integrated organism. Jack Sail-by-the-Wind *Velella* is much smaller than the Portuguese man-o'-war and has a flat, oval float bearing a sail which projects across the float at an angle. *Velella* has a single feeding polyp under the float surrounded by reproductive polyps and the whole colony is fringed with stinging polyps. The stinging tentacles of the latter are much shorter than those of the Portuguese man-o'-war and the animal is not at all dangerous to man. It feeds mainly on small shrimps and other planktonic animals. The fact that different strains of *Velella* have the sail running in different directions across the float to aid dispersal by the wind has already been mentioned.

Sea Mats. These small animals often called bryozoans or 'Moss animals' are almost entirely colonial. Most of them occur in the sea but they are also found in fresh water. The individuals themselves are microscopic but form sometimes large colonies that may reach 8 in (20 cm) in size. Each individual zooid resembles a cnidarian polyp superficially, but they are very different in their internal structure. The mouth of the zooid is surrounded by feeding tentacles but the gut is U-shaped and has an anus. There is also a true body cavity. Each zooid lives in a box-like skeletal capsule and has special muscles that attach it to the walls of its box-like home. Eversion is brought about by devices like collapsing lids and walls that increase the pressure inside the capsules so that the zooids are popped out. Only some 'Moss animal' zooids are actually in cytoplasmic continuity with their neighbours, others merely touch one another externally; nevertheless the members of these colonies can work in co-operation. *Membranipora* is a very common bryozoan often found encrusting rocks and sea weeds on the shore. It forms flat lacy colonies that may have as many as 2 million members. This colony extends over many square inches but many bryozoans form smaller colonies than this. The hornwrack *Flustra* is exceptional in size and can measure over 6 in (15 cm) in height and forms large stiff fronds like those of sea weeds. Its buff grey fronds are often found washed up on the beach and close examination shows them to be made up of many minute capsules packed closely side by side. In a 'Moss animal' called *Bugula* which consists of bushy tufts of branches and grows under rocky overhangs on the middleshore there are specially modified zooids called avicularia which have a beak-like structure and prevent organisms settling on the colony. These look remarkably like the head of a bird and have a function very similar to the pincer-like pedicellaria on the skin of Sea stars and Sea urchins. One surprising bryozoan is a fresh water form called *Cristatella*. This consists of a greenish gelatinous colony which may reach 8 in (20 cm) in length. Unlike most other 'Moss animals', *Cristatella* can crawl along on its

A colony of common Sea squirts *Halocynthia papillosa* showing the two siphons through which water currents enter and leave the body.

Although it resembles a seaweed this hornwrack is made up of thousands of box-like capsules each containing a living moss animal. It is in fact a colony of animals, not plants!

flat foot-like region. The zooids with their feeding tentacles are all situated on the top of the colony.

Sea Squirts and Salps. The Sea squirts are filter feeding animals that look rather jelly-like but in fact are enclosed in a stiff tunic that protects their fragile bodies. They have two conspicuous siphons on the body, an inhalent siphon at the top through which water enters and an exhalent siphon at the side through which it leaves. Inside the body is a perforated pharynx consisting of a mesh-like straining device. Water is drawn into the body by cilia and strained through the pharynx which is coated with mucus that traps food. The food particles are collected together and are moved down into the gut for digestion in the form of a food string. The gut is bent round in a U shape so that waste is emptied into the stream of water passing out through the exhalent siphon. Sea squirts encrust rocky overhangs and sea weeds on the shore and sometimes foul the bottoms of ships and pier pilings. The Group is entirely marine and Sea squirts may be individual or colonial. A common European Sea squirt *Clavelina* about 1½–2 in (4–5 cm) tall forms clumps of tube-like semi-transparent colonies. The individual Sea squirts produce long stolons or rootlets and new individuals are budded from these. The rootlets hold the individuals together and anchor them to rocks.

The compound Sea squirts form much more intimate colonies. These are the beautiful daisy-starred encrusting organisms often found on sea weeds and rocks. The individuals have separate inhalent openings but share a central exhalent opening and are grouped around this, each tiny Sea squirt looking like the petal of a flower. The individuals may be brown, yellow or blue and are grouped around an orange exhalent siphon at the centre. There are many separate daisy-shaped colonies scattered at intervals throughout a common, enveloping jelly film which is quite firm so the

Sea squirt colonies on the shore arise by budding. Each Sea squirt produces a rootlet which buds off new individuals. In this case the individuals are morphologically, as well as genetically, identical.

whole looks like a gorgeously embroidered mantle. The details of these colonies are best seen with a hand lens as the individuals are very small. In *Botrylloides* the individuals are not arranged in a daisy pattern but in two parallel rows to form a pattern wriggling along under the surface of the protective jelly film. In this colony the small Sea squirts share a common exhalent canal that opens by a pore at intervals.

The salps are transparent, barrel-shaped tunicates that float in the surface waters of the sea. They are ringed with muscle bands and have openings fore and aft. Water is taken in through the front opening and passes through the pharynx slits leaving small food organisms trapped in the gut where they are digested. The water is expelled through the rear opening by contraction of the muscle rings and serves as a kind of jet propulsion. These drifting individuals produce long trailing stolons which bud off miniature adults. These can remain attached in a kind of train for some time. Salps can also reproduce sexually.

Invertebrate Colours and Venoms

Tropical invertebrates of all kinds show a richness and exuberance in colour and pattern that easily rivals that of fish and birds. In warm seas the coral reef and its inhabitants blossom with all the brilliance of an exotic underwater garden. The coral terraces are spangled with Velvet sea stars of crimson, indigo and orange and plumed with Sea lilies of scarlet and chrome yellow. Venomous Cone shells are blotched with warning cream and sepia or etched with a network of brown on gold whilst cowries stretch gaudy mantles of living tissue over their smooth shells. The Sea slugs are amongst the most gorgeous of the reef's inhabitants and have frilled and tasselled backs of every conceivable colour combination. Polyclad flatworms, fanworms, crabs, lobsters, shrimps and octopuses are also dazzlingly patterned. On land are the tropical butterflies, the eyed, barred and tufted caterpillars, metallic beetles, patterned bugs and shimmering dragonflies. There are banded snails, blue and red Coconut crabs and striped flatworms and leeches. These intense colour patterns are largely reserved for the tropical seas and rain forests where light and shade are both intense and the warm climate supports a vaster number of competing animals than in temperate climes. Where great numbers of animals occur close together sporting a distinct livery

for intraspecific recognition, warning or camouflage is obviously important and, as we shall see, animals adopt bright colours for many reasons. More exposed tropical situations like savanna and desert support fewer invertebrates since apart from insects these are in the main very much dependent on water or a humid environment. Those that do occur in these and other regions are often cryptically coloured in buffs, beige and brown. The invertebrates of temperate climes are drab in comparison with their rain forest or reef living relations but this enables them in general to blend in better with the more sombre colours that surround them.

Cryptic colouration is one of the most important aspects of animal colour. By having colours that blend with the background with a disruptive pattern and countershading that will break up the body outlines, the motionless animal can merge into its surroundings and become almost indistinguishable from them. This is useful not only for animals hiding from predators but also for the predators themselves that lurk unseen in wait for their prey. As we have seen some invertebrates such as cuttlefish, octopuses and squid have a colour change system under nervous control so they can rapidly adapt their cryptic colouration to suit their background and circumstances. Crustaceans can change

The Sea slugs are some of the most exquisitely patterned of marine animals. The gold veined rosette on this one is a circlet of gills.

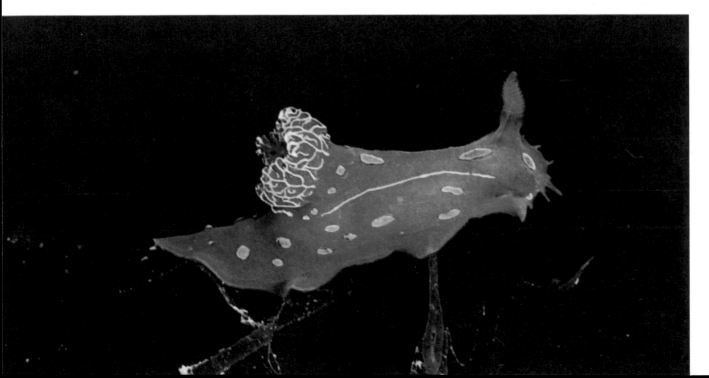

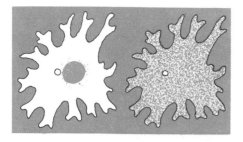

Two ways in which colour change can be effected. (Above) The pigment cell of a vertebrate with the pigment concentrated into an inconspicuous ball (left) or spread out and thus colouring the cell (right). (Below) The pigment cells of a cuttlefish showing has contraction of the ray-like muscle fibres spreads the pigment from a small globule (left) to a large and more conspicuous disc (right).

their body colour more slowly, this being under hormonal rather than direct nervous control. Animals like the Sea hare change their colour according to the kind of sea weed they are eating and so become very difficult to see amongst the red or green fronds of their food plant. The shell of the Dog whelk becomes coloured differently according to diet, but this change is irreversible once the shell has been formed. Dog whelks feeding on barnacles have white shells and those feeding on mussels have brown or mauve shells. Other invertebrates such as the Banded snail *Cepaea* have genetically determined colour patterns in their shells that help to camouflage them in different surroundings. Banded snails are polymorphic for shell pattern and colour and particular colour patterns have been shown to have an adaptive significance in different habitats, although the situation is far from simple. Snails with plain yellow shells seem to survive best in green surroundings. Those with yellow shells marked with dark stripes survive best in tall grasses and hedgerows and those with plain pink and brown shells survive best in beech woods. These latter colours also offer the best protection early in the year. The snails are preyed upon by thrushes which batter the snails on hard ground or on special stones called anvils to crack open the shell. Examination of thrush's anvils for shell fragments will

◁ The Tiger cowrie *Cypraea tigris* in life shows the patterned shell partially overlain by the tentacle-fringed mantle of living skin.

This gorgeous Sea slug has colourful tassels called ▷ cerata on its back. In some Sea slugs the cerata contain stinging cells, carefully harboured from corals, hydroids or similar prey, which seem to be functional even though they have passed through the Sea slugs gut. They probably deter would-be predators, an important defense in a shell-less animal.

show what kind of snail the thrush is finding in different habitats and therefore which kinds of snails are less well adapted to these habitats. Another kind of camouflage involves producing secretions that act like a smoke screen to confuse predators or mask a rapid escape. Octopuses, cuttlefish and squid discharge an inky solution containing melanin into the water when they are disturbed and may at the same time blanch white and shoot off in another direction as it were under cover of darkness. The Sea hare *Aplysia* emits a purple pigment, apolysiopupurin, into sea water when irritated. In direct contrast to animals using colour to conceal themselves are those that display bold colour patterns to advertise the fact they are poisonous and best left alone. Many tropical Sea slugs employ warning colours because although they lack a shell and might appear to form a 'bonne bouche' for a predator, they often have, in fact, the custom of feeding on Sea anemones, jellyfish and colonial hydroids. From their prey they acquire stinging cells which are not digested but are passed

undischarged into the most superficial thin walled branches of the gut. From here they can be used as a defence against inquisitive animals or predators. One Sea slug advertises the fact that it contains stinging cells in a very direct way. The British aeolid, *Aeolidia papillosa* which feeds on the Beadlet anemone, has long processes on its back that resemble the tentacles of its prey and it seems to be mimicking its prey. The phenomenon of visual mimicry is widespread in invertebrates but has been best studied in insects.

Colour patterns are also used as intraspecific recognition signals, for instance during courtship in advanced invertebrates such as cephalopods. Sometimes the colour of an animal is not self determined but is being manipulated by a parasite it contains so that parasitized members of the population may stand out from the others in a most conspicuous way. Parasites that make their intermediate hosts more conspicuous to predators may further their own ends, because if the predator is the final host, the parasite is ensuring it gets eaten.

Water fleas *Daphnia* in stagnant water can increase their haemoglobin content ten times. This gives them a pinkish appearance and helps them to obtain oxygen from the water.

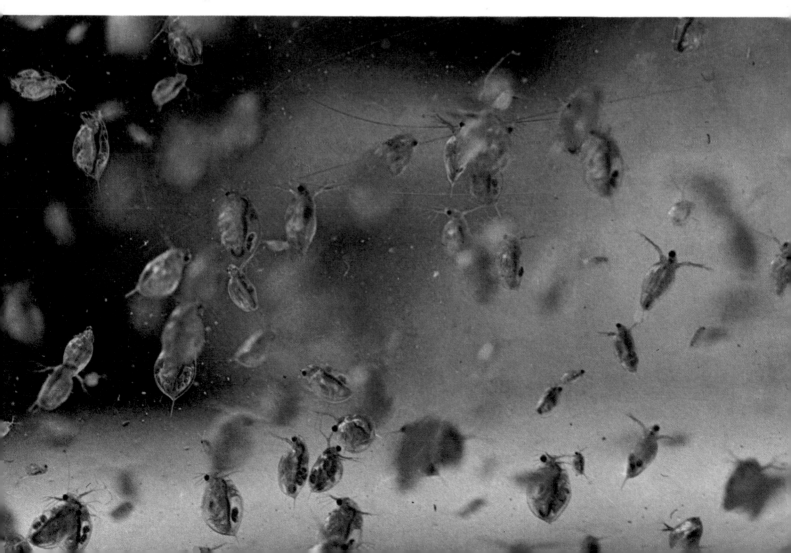

Some parasites alter the colour of their hosts by producing metabolic disturbances, and where the change in colour produced affects a final host it is unlikely to further transmission of the parasite. The Water flea *Daphnia* turns red when infected by a certain bacterium that produces the pigment rhodo-violacin. The marine copepod shrimp *Calanus* turns orange-red when infected with protozoans in the body cavity and *Sacculina* parasitizing crabs can make them look redder than usual. Algal symbionts inside the tissues of Acoel flatworms, Sea anemones and hydras can colour them bright green. Environmental factors can also influence animal colour. Some deep water and cave-living invertebrates are albinos because light is needed to make the black pigment melanin. Several flatworms living in alpine lakes exist as dark forms in the shallow water at the edges and pale forms in deep water. Temperature can also affect the deposition of melanin in some arthropods. Arctic forms of Water flea, *Daphnia*, are darkly pigmented on their backs and some insects reared at low temperatures become melanistic.

Physiological Pigments. Invertebrates also use pigments for physiological purposes. Many, but by no means all, invertebrates have coloured pigments that carry oxygen to the tissues or act as an oxygen store. These pigments may occur in special blood corpuscles, may merely be dissolved in the blood plasma, or may occur in the muscles and other tissues. Not all invertebrates, of course, have a blood system but they may have respiratory pigments. Haemoglobin is an iron-containing pigment that gives the blood of humans and other vertebrates its red colour. It is scarlet when oxygenated and purple when deoxygenated. It is widespread throughout the animal kingdom and in invertebrates occurs in the ciliate *Paramecium*, in the parasitic roundworm *Ascaris*, in parasitic flukes such as Liver fluke, in *Tubifex* worms, in ragworms, lugworms and earthworms, in *Daphnia*, in the 'bloodworm' larvae of the midge *Chironomus* and in Water boatmen. Molluscs with haemoglobin are the Coat of mail shell, *Chiton*, limpets, whelks, the Sea hare and some pond snails such as the Ramshorn snail *Planorbis*. A few Sea cucumbers also have haemoglobin. Several of these animals can increase the oxygen carrying capacity of their tissue fluids by synthesizing more haemoglobin if they live in oxygen deficient water. This is well known in the Water flea *Daphnia* which can increase its haemoglobin content tenfold in poorly oxygenated

A painted Top shell from the shore owes its pink shell markings to substances called porphyrins.

139

The Sea mouse Aphrodite is an unusual polychaete worm and has iridescent chaetae along the edges of its body. The total length of the worm is about 8 in (20 cm).

water and may change its colour from almost colourless to bright pink. A swarm of *Daphnia* in a poorly aerated pond can make the water look bright red. Young Ramshorn snails *Planorbis* and bloodworms can also increase their haemoglobin content in this way. Another iron containing pigment which is completely unrelated in structure to haemoglobin is called haemerythrin. This is red when oxygenated and pale yellow when deoxygenated. It occurs in Lamp shells *Lingula* and in Siphunculid and Priapulid worms. Animals such as cuttlefish, octopus, the Garden snail, lobsters, King crabs, scorpions and spiders have blue blood due to the presence of a copper containing blood pigment called haemocyanin. This pigment is only blue when oxygenated and is almost colourless when deoxygenated. Copper is a poisonous metal and it is interesting that even minute amounts dissolved in water will kill animals that have an enormous concentration inside as blood pigment. The copper only becomes innocuous in organic combination with other molecules; haemocyanin itself is a protein copper compound. Possession of haemocyanin as a blood pigment may have limited the success of cephalopod molluscs which at one time almost dominated the seas as the fossil skeletons of extinct ammonites and belemnites testify. This group which had such a phenomenal success in the sea, which is rich in copper washed down by rivers, have never invaded the land. Could this be because they have committed themselves to respiration involving gills which would be useless on land; have lost their shells and could not support themselves out of water or is it that they would not find on land sufficient copper to make the enormous quantities of haemocyanin they need? Certainly Garden snails and some pond snails manage to concentrate copper from their food enormously but these are much smaller animals than many cephalopods. A few polychaete worms such as the Peacock worm *Sabella* have green blood and this is because they contain the green pigment chlorocruorin which despite its colour is an iron containing pigment closely related to haemoglobin. In fact when concentrated, blood containing chlorocruorin looks red but in dilute solution it appears green. This means that large blood vessels may appear red and small ones green in the same animal.

Related to haemoglobin and chlorocruorin are the porphyrin pigments. These only occasionally exist in a free form to give colour to invertebrates, yet where they are found, they are of great interest. The purple colour of the dorsal cuticle of the earthworm is due to free coproporphyrins and these are also said to occur in the skins of Garden slugs and the Sea star *Asterias rubens*. A high free porphyrin concentration in man is associated with the often fatal congenital disease porphyria. One of the strange properties of porphyrins is that they are photodynamic. That is, they absorb light, particularly in the ultra violet, and photosensitize the tissue containing them causing metabolic disturbances

The white cross on the back of the female Garden spider is produced by guanine crystals.

◁ A scorpion from the Isle of Elba. Most European scorpions do not have a very dangerous sting.

A Black widow spider of North ▷ America, one of the few spiders dangerous to man. Its bite while extremely painful is fatal in a small percentage of cases only.

leading to complete breakdown of the tissues. In slugs and Sea stars the free porphyrins seem to be masked from light by other pigments but earthworms are not so protected and they can be killed by exposure to light. Oddly enough, free porphyrins account for the red markings on the shells of cowries, chitons and Top shells and have also been identified in the shells of fossil molluscs by the brilliant red fluorescence they give in ultra violet light.

Few animals apart from *Daphnia*, *Tubifex* and Blood worms owe their general body colouration to their blood pigment. Usually the skin contains special pigment cells that give them their colour. Other pigments that invertebrates use physiologically are the visual pigments that occur in the light receptive cells in their eyes and screening pigments that line eye cups or separate the ommatidia of compound eyes to prevent cross illumination. Other uses of screening pigments are to serve as a

The Giant scoloperdra centipede of Trinidad reaches over 10·25 in (26 cm) long and is ten times larger than most centipedes. They have a painful though not fatal bite.

shutter to screen light emitting organs, or to protect symbiotic algae that are being formed in the tissues of an animal, for example, the Giant clam. The freshwater tropical snail *Pomacea canaliculata* lays masses of rose red eggs out of water. The colour is due to a carotenoprotein which is extremely resistant to heat and may therefore serve as a heat screen to protect the delicate developing embryos in the eggs. The protein alone with the carotenoid pigment split off from it is very unstable to heat.

Animal Colours. The main pigments producing colouration in animals are the black pigment melanin, and chemicals called ommochromes, carotenoids, pterines and guanine. Melanin is the dark pigment that occurs in human hair and gives sun tanned skin its bronzed colour. In invertebrates it occurs in the skin of slugs, in some insect cuticles and gives the ink of cuttlefish, octopus and squid its colour. The ink of the cuttlefish *Sepia* is used to produce artists' sepia pigment and is still marketed. Fossil ink sacs containing undischarged ink have been found with squid remains on the South coast of England at Lyme Regis. The melanin suspension had survived 150 million years and was used by its discoverer to illustrate the monograph describing the squid that produced it.

Not all brown or black pigments are melanins. The black pigment in the ommatidia of insects and crustaceans is an ommochrome. Related to the ommochromes is a remarkable pigment called tyrian purple produced by Mediterranean whelks *Murex* and the Dog whelk *Nucella lapillus*. The pigment is produced by a mucus gland in the mantle cavity and is at first cream but undergoes a photochemical change through green to purple in sun-

light. The precise use of this secretion to the whelk is not known but it may have an unpleasant taste and deter predators. Cloth has been dyed with the pigment extracted from *Murex* shells since Egyptian and Roman times and the name tyrian purple comes from the fact that the Phoenecians developed the dyeing industry in Roman times. Wearing of tyrian purple was a sign of rank in Ancient Rome. The dye is extremely stable once the purple colour has matured and will not fade in sunlight.

The carotenoid pigments are usually orange, brown, yellow and red but can be blue. They are entirely derived from plant pigments taken in food and cannot be synthesized by the animals themselves. As vitamin A, carotenoids are the source of the visual pigments. Many crustaceans owe their colouration to astaxanthins which are carotenoid pigments. The red antennae and blue-black body colour of living lobsters are due to these pigments. When lobsters and prawns are cooked they become bright pink because heat denatures or 'kills' the proteins linked to astaxanthin and liberates the pigment which becomes oxidized to bright pink astacin. The green colour of shore crabs and the blue colour of Jack Sail-by-the-Wind which feeds on planktonic crustaceans are both produced by carotenoproteins, as are the yellow, red, brown and violet colours of certain Sea stars. The fact that

cockles and mussels can store carotenoids from the algae they feed on is used by many zoos which feeds them to the flamingoes in order to keep their feathers in the pink, since their feather pigments are also carotenoids.

Pterins are white, yellow, orange and red pigments. Some of them are related to uric acid and in fact the wings of the Cabbage white butterflies are coloured by an excretory product that has found a new use. The yellow colour of wasps is due to a pterin and the yellow chromatophores of certain crustaceans also owe their colour to these pigments.

Guanine is an excretory substance but is also a component of nucleic acid, so is present in all cells. In some tissues it accumulates as minute granules and crystals and when the crystals are arranged in regular arrays they form very good reflectors and can give animals a silvery appearance. The silvery colour of fish scales is in fact due to the regularly arranged guanine crystals they contain. Cephalopods have iridescent 'chromatophores' in their skin called iridiocytes which owe their colours to physical colour produced by guanine crystals. The tapetum, or reflecting layer at the back of the eye of the scallop, which gives the eye its steely blue colour, is also due to guanine crystals. Guanine granules can also produce a mat white effect as in the white cross on the back of the Garden spider.

143

Animal colours are not always caused by pigments but may be structural colours produced by surface texture. The iridescence of the iridiocytes of the cuttlefish and shifting blue-purple colour of South American *Morpho* butterflies are examples of structural colours. The colours of the latter are produced by light being reflected from the ridges and grooves of a number of lamellae covering each scale on the wings. The gold-green iridescence of the chaetae surrounding the body of the Sea mouse earn this worm its name, *Aphrodite*. This iridescence is produced by diffraction of light from chitin fibres in the chaetae and disappears if the chaetae are dried out. Apart from this, its crowning beauty, the Sea mouse is a fairly drab animal.

Venomous Invertebrates. Many groups of invertebrates have independently evolved venoms as a means of catching and paralysing prey, as a means of self defence or for both purposes. Venoms are not only directed against other species. In the social insects in particular, venoms may be used against members of the same species. In the case of bees, if more than one young queen is reared they will sting one another to death until only one remains. Man has long known those animals that have poisonous stings or bites and different cultures have alternately revered or loathed them. Here mention will be made only of those invertebrates that are venomous to man. There are three main kinds of toxins produced by animals. Irritants produce a local tissue response or bad stink. Haemolytic substances cause breakdown of blood cells and tissues and neurotoxic substances act as nerve blocks and cause paralysis often culminating in respiratory failure and death by suffocation. Most of these toxins have to be injected into the body to produce an effect. If a predator eats a venomous animal the venoms will be broken down in the gut and rendered harmless.

Apart from insects, the spiders, scorpions, centipedes and millipedes all have poisonous representatives. Spiders have poison fangs which they jab into their victims to inject venom. One of the most deadly spiders commonly encountered by man is the Sydney Funnel web spider. This is a large spider that occurs in some of Australia's most densely populated areas. It lurks in gardens and under the porches and foundations of houses and many of its victims are children walking around bare-footed and having less resistance to the venom than adults. Control experts are employed by local governments to catch the spiders and spray the areas around houses against them. The venom of the female spider is not as poisonous as that of the male, and fortunately it is the female spider that is encountered most frequently. Research is presently in progress to develop an antivenine to this neurotoxin. The Black widow spider, which occurs throughout the USA has a much worse reputation for being dangerous than it in fact deserves. The Black widow female is a small, shiny black spider with a red hour-glass mark under the abdomen. It gets its name from the fact that it is reputed to devour the male after courtship although this does not invariably occur. Only the female spider is at all dangerous to man since the male is too small to pack enough venom to do much harm. The female spider is again common around human habitations and is apparently often encountered in outside latrines. The form in the southern States is more poisonous than the northern form and injects a neurotoxin that can be serious, causing pain, cramp and paralysis but is rarely fatal. Deaths from authenticated Black widow attacks run at about 1 every 4 years in the USA. It is quite likely that shock is a contributory cause of death in these cases.

The hairy tropical Bird-eating spiders which can reach a handspan across cause a revulsion in the beholder out of all proportion to the venomous effect they can produce. Although they can kill and eat humming birds, their powerful fangs inject a substance that may cause a reaction no worse than a bee sting in man. It is not clear whether the fluid injected contains only protein splitting digestive substances or whether true venoms are also present. The cuticular hairs of the spider are probably more dangerous than its bite. They break off in the skin if the spider is handled and produce painful swelling and irritation.

Scorpions, unlike spiders, have a sting in their tail which is used both in defence and attack. Most species of scorpion are fairly harmless to man and in those that can kill size is not related to dangerousness. The sting of some of the commonest scorpions in southern Europe produces only slight pain and is relatively harmless. The small brown Buthid scorpions of southern Europe and Africa, however, have a deadly nerve poison equivalent in virulence to that of a cobra that can cause death with convulsions and frothing at the mouth within a few hours. Despite their powerful deterrent scorpions preyed upon by other animals including baboons which break off the tails before eating them. Centipedes are carnivores that have adapted the

Large hairy spider of the tropics popularly called a tarantula eats small birds but its bite is not particularly dangerous to man. The hairs, however, may have a powerfully irritant action.

first pair of legs as poison claws. Some tropical centipedes grow very large and the Giant Malayan centipede that grows to 8 in (20 cm) long has been known to kill man. The *Scolopendra* centipedes of South America can grow up to 1 ft (30 cm) long and have been fed on mice in captivity although they eat locusts, cockroaches and geckos in the wild. Their bite rarely proves fatal to man but is excessively painful. Millipedes are herbivores and so do not need poison claws for catching food. They do however have poison glands down the sides of their body

for defence and produce noxious secretions from these. One poisonous centipede from central Mexico is ground up and used along with various plant extracts as an arrow tip poison. Usually millipedes only ooze venom, but a few tropical forms can fire a fine spray of irritant up to a distance of a yard. This has been known to blind curious birds and cause human skin to blacken and peel off. Some of the Flat-backed millipedes produce hydrocyanic acid which vapourizes into cyanide gas, to repel enemies. The secretion is 'made to order' by the

145

mixing of the contents of two glands usually kept sealed from one another by a sphincter muscle.

Poisonous Marine Invertebrates. The sting cells of the cnidarians have been mentioned several times. Only a few cnidarians are seriously poisonous to man. The stings of the Portuguese man-o'-war can cause severe irritation lasting for several days to swimmers but is very rarely fatal. One of the most deadly animals known are the Sea wasps or Box jellies which are jellyfish with a cube-shaped bell and clusters of stinging tentacles hanging at the four corners. These come inshore in large swarms to northern Australian beaches at the hottest time of year in calm weather when bathers are most likely to be about. The larger Sea wasps can reach 10 in (25 cm) in length of the bell but the tentacles can extend for 30 ft (9 m) and are very difficult to see in the water because of their transparency. Each tentacle is armed with millions of stinging cells containing one of the deadliest venoms known. Toxin extracted from them and diluted 10,000 times can still kill laboratory animals. The venom is probably a simple protein. That of *Chironex fleckeri*, one of the deadliest Sea wasps, causes excruciating pain and in high concentrations can cause tissue lesions followed by collapse and sudden death. The venom usually only proves fatal if man receives a high dose, nevertheless it has caused several deaths. Attempts to protect beaches with fine mesh nets or to develop a repellent have failed and beaches may fly warning flags where the Sea wasp invades. An antivenine has, however, recently been developed.

The Cone shells of tropical waters are molluscs which have become collectors' items owing to their beautifully patterned shells. The Glory of the sea *Conus gloriamaris* has been one of the most prized sea shells for about 200 years and the record price of $2,000 was paid for a single shell in 1964. Some of these cones are, however, exceedingly poisonous to man, particularly those that feed on fish; those feeding on other molluscs and on worms are less dangerous. The venom of the cone is injected through a harpoon-shaped hollow tooth on the 'tongue' or radula of the molluscs. Cones preying on fish unsheath a long feeding proboscis and from this one of the teeth is discharged like a harpoon into the prey. The poison injected through the hollow tooth causes paralysis and death within a few seconds. The radula contains several teeth and is continuously formed and a new tooth is used for each victim. The toxin is a nerve poison and causes muscular paralysis. It consists of several substances including amines and proteins. Why a Cone shell wishing to capture a small fish should inject sufficient poison to kill a man is a matter for speculation. It may be that man is particularly sensitive to the toxin. Cones are known to have caused death in 20 per cent of the cases that man has been 'bitten', a record that rivals that of some of the deadlier snakes. The molluscs are obviously very able to abet their own conservation. Another potentially dangerous mollusc, though a comparatively little studied one, is the Blue ringed octopus of Australia. This small octopus is beautifully patterned with black, yellow and blue bands that extend over the arms as well as the body. It is known to have caused only one death by biting a swimmer on the back,

Richly coloured crimson and white Cushion star.

Cup corals *Caryophyllia clavus* are small British corals. The mysterious green colour of the polyp is probably an interference colour.

but a single case is difficult to evaluate as the subject may have been hypersensitive to the nerve toxin injected. A marine animal that affects man in temperate waters is the bryozoan *Alcyonidium*. These organisms form colonies of 4–8 in (10–20 cm) long with the texture of firm jelly. They are common in the North Sea and are often caught in trawl nets. Contact with these colonies or with nets and ropes that have been in contact with then can produce a severe dermatitis with a painful rash and blistering. The condition is known popularly as Dogger Bank itch.

Amongst the echinoderms several Sea urchins bear pincer-like pedicellariae which secrete toxins. Their function is actually to kill and remove organisms that may settle on the skin of the Sea urchin, but some pedicellaria can pierce the skin of man and the toxin emitted is said to cause pain, dizziness and even facial paralysis, even though only a finger has been punctured. The toxin doubtless also enters through wounds made by spines or other abrasions. The hollow spines of echinothurid Sea urchins are themselves poisonous and have a poison sac at their base. This may be true too of the notorious Crown of thorns starfish which avidly devours living corals in many warm seas. The spines of this monster starfish are said to produce painful wounds. Sea cucumbers lack spines and have a reduced skeleton, and so rely heavily on behavioural and chemical defence mechanisms. The evisceration technique of Sea cucumbers has already been mentioned, also the discharge of noxious yellow mucus threads by the Cotton spinner. The general body wall of Sea cucumbers may also discharge a secretion that is toxic to certain fish. Pacific islanders use the toxins of Sea cucumbers as a means of stupefying fish in rock pools. The Sea cucumber is macerated and liberates a toxin called holothurin A that has haemolytic properties and a neurotoxin that also has antibacterial and antiprotozoal properties and might be used to prevent settlement of encrusting animals as well as to repel predators. The haemolytic factor affects the fish via the gills but is only slowly absorbed so that the fish are not themselves toxic after capture by this method. Dried Sea cucumber known as beche de mer or trepang is itself eaten as a delicacy in parts of the world, and is therefore not poisonous so any toxins present must be either unstable and not survive the drying process or be easily digested in the gut.

Man and Invertebrates

As 95 per cent of all living animals are invertebrates it is hardly surprising that they exert an enormous influence on the life of man. Invertebrate animals are so numerous and so diverse that they form the basic fabric of life. Many enter into food chains in which man is directly involved such as those where man's food fish have themselves fed on planktonic or bottom living invertebrates. Man himself also eats invertebrates directly and some constitute a great delicacy, this according very much with habit and custom. Succulent crustaceans prized as food include shrimps, prawns, scampi, crab, crayfish and lobster. The molluscs are another good group for the gourmet and furnish us with the delectable cockles, scallops, clams, mussels, winkles, whelks, oyster, squid, octopus and abalone. This last mollusc is a kind of limpet with an ear-shaped shell punctured by a whole series of holes along one side. The broad muscular foot is the part that is eaten and is encountered as awabe in Chinese restaurants. The abalone occur in the Channel Islands where there is a great demand for them. The Roman or Edible snail (escargot) is eaten in France and elsewhere in Europe but is only slowly regaining favour in Britain. There is evidence of a long snail-eating tradition in Britain, however, for snail shells found in Bronze Age barrows suggest that they were eaten in 1800 BC. Later the Romans used to cultivate snails in 'cochlearia' and may have brought this habit to the West of England. In the 19th century 'wall fish', as snails were then called, were widely eaten and were sold in Covent Garden and elsewhere. Snails have also been imputed with healing powers, particularly against coughs and colds but also against corns. The slime of the snails was used to produce a greenish salve for corns and as late as 1880 plasters made from thin paper over which snails had been allowed to crawl were sold in London. Oysters have been cultivated as a delicacy since Roman times and different cultures use their own particular methods of inducing the delicate oyster spat to settle. The Romans made good advantage of the indigenous oyster beds they found around Colchester and elsewhere when they arrived in England and consumed large quantities of them. They also exported British oysters to Rome. During the last century oysters were a poor man's dish and 40 million a year were being sold. This led to their

The European lobster. This and other crustaceans have long been prized as food.

Mytilus edulis the Common mussel anchored by its byssus threads. The byssus threads of the much larger Fan mussels were once woven into cloth of gold.

becoming badly overfished, and they have also been run down due to pollution and other factors so they are now in short supply and expensive. Green oysters are particularly prized. The colour is produced by algae that have been strained out of the water on to the oyster's gills as they feed. British cockle beds are also at present in danger and in South Wales there is a misguided move to cull large numbers of Oyster catchers (which do in fact eat cockles and a variety of shell fish) in a vain attempt to preserve the cockle beds. More unusual sea fare is provided by the echinoderms. Dried Sea cucumber is eaten in various parts of the world, particularly by the Japanese, Chinese and Maoris. It is reputedly somewhat leathery but not unwholesome. The roe of Sea urchins is eaten by Mediterranean peoples, often on toast, and might one day form an alternative to smoked cod's roe and caviar. As yet the charm of eating insects has not made itself widely felt in the West, although small pockets of devotees have appeared. This is largely due to a drive launched by the Japanese, a nation noted for its gustatory enterprise, to expose the West to tinned butterflies' eggs, caterpillars, and chocolate coated ants. Again, Roman concupiscence knew of insect delicacies and epicures fattened Stag beetle larvae with flour and wine to be eaten as 'cossus' at banquets. Aborigines are very partial to their grub whilst Hottentots favour caterpillars. Darwin himself tried eating the larva of the Sphinx moth which he voted delicious. John the Baptist is reputed to have eaten locusts, although there is dispute about whether the locusts involved were not the beans of the cassia or locust tree. Inhabitants of Arabia, Madagascar and Africa reputedly eat fried locusts after stripping off the legs and wings. The Manna with which the Israelites were miraculously provided may be more prosaic than it seems, for it has been suggested it may be the sugary excreta of aphids which coats the leaves on which those insects have been feeding. Another insect product greatly favoured by man is of course honey. Spiders seem unpromising fare, but a spider hidden inside a raisin was prescribed as a cure for malaria by the famous Dr. Muffet whose daughter became the subject of a well known nursery rhyme.

Pearls. Invertebrate animals also feature in human activities as diverse as magic and economics. Many are used either directly or indirectly in man's personal adornment and the molluscs in particular have contributed much to fashion over the ages. Pearls which though today are produced mainly by the Pearl oysters *Pinctada partensi* can also be produced by other bivalves. In Roman times the freshwater Pearl mussel *Margaritifera margaritifera* which occurs in soft waters throughout Britain was

Earthworms play a major role in aerating and turning over the soil.

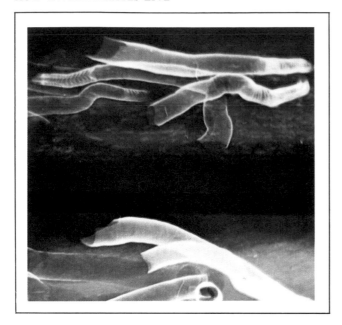

An X-ray picture of wood infested with shipworm. The calcium lined tunnels show up in the photograph. Ship worms are a considerable pest as fouling organisms and may have played a part in the defeat of the Spanish Armada.

Acanthaster planci, one of the sunstars is responsible for ▷ laying waste large expanses of coral reef as it eats living coral. A natural predator for this starfish is being sought in order that it may be controlled.

the basis of the British pearl industry. Pliny was well aware of the dangers of pearl fishing and warned that one's fingers were in danger when living mussels were probed for pearls. The pearls gained by taking such risks were often made into earrings for Roman matrons. Pearls are produced naturally in response to small irritant sand grains or grit embedding themselves in the living mantle of bivalves. Plates of pearly nacre are laid down around the particle which becomes encapsulated in the centre. Cultured pearls are produced by inserting a small glass bead inside the oyster and then waiting for several years until the pearl is fully formed. The colours of pearls range from white, cream-yellow, rose to black. Black pearls are the rarest and most valuable. Prodigiously large pearls can be formed in Giant clams and the largest of these called the Pearl of Allah weighed 14 lb (6 kg) and measured about 10 by 5 inches (25 by 12·5 cm). Most of the pearls produced in these Giant clams are valueless, yet they are investigated for pearls by divers and there are apocryphal stories about divers being killed by the Giant clam closing its valves on the diver's leg and holding him underwater until he suffocates. The shell of the Pearl oyster is also valuable as mother of pearl used in inlay work. The abalone or Ormer shell is also beautifully iridescent inside and at one time abalone mother of pearl was used to provide the inlay decoration on violin bows. In the Channel Islands, islanders have decorated houses with the Ormer shells. Abalone shells were once also used in the button industry, the buttons being stamped from the shells. Top shells were also used in the button industry at one time and produced red, porphyrin-mottled buttons. The plastics industry is putting these uses out of fashion however. The Glossy cowries have long been used in decoration and magic, notably by Africans and islanders in the Indian Ocean and South West Pacific. The magic properties associated with cowries may be due to the Ionic appearance of the shell's aperture. The generic name *Cypraea* of the cowrie is said to come from Cyprus where there was a flourishing cult of the goddess Aphrodite. Like the Cone shells mentioned in the last chapter, cowries are prized as collectors' items and some such as the Golden cowrie may fetch large prices. The Money cowrie has been used as currency from very early times, and in the coastal areas of West Africa were usually threaded onto strings in 40's or 100's. About a hundred years ago, 50 strings of 40 were worth an American dollar. The shells were frequently imported from the Indian ocean area via England, which re-exported them and some 60 tons of Money

Cowries, with one seen in section to show the spire. These shells have long been associated with magic and some are also collector's items.

Murex tenuispina from the Indian ocean is related to the murex snails in the Mediterranean that furnished the famed Tyrian purple used as a dye by the Phoenicians and Romans.

cowrie were said to have passed through Liverpool docks in 1848. Cameos are made from whelk shells or tropical Helmet and Bonnet shells by cutting the opaque surface layers into patterns which stand out in relief against the translucent honey or brown coloured inner layers of the shell. Until recently, sailors have produced decorative scrimshaw work by engraving designs onto shells and whale's teeth. The Indian Chank shell is a large conch that has become the centre of much superstitious custom and is used as a trumpet at Hindu marriage services and on other religious occasions. It is also cut into bangles of various kinds which are usually lacquered. Some lacquer comes from a scale insect

which is crushed to form the resin. The shell of the Pearly nautilus, a cephalopod, used to be mounted in gold and silver and made into gorgeous chalices and drinking vessels and more prosaically the chalky skeleton of cuttlefish is given to cage birds. Other practical uses found for shells are as water carriers. The False trumpet, a tropical snail, grows to 20 in (50 cm) long and is the largest of the snails. Its shell is used as a water carrier by natives of North West Australia. The honey-coloured Baler shells were used to bale out canoes. An unusual industry based on molluscs was the weaving of cloth from the byssus threads, or beard, of the Fan mussel. This industry was centred around Taranto in Italy, and gloves and other articles made in this way can still be seen in museums as they are extremely durable. The byssus threads are made of tanned protein and have a golden sheen so that the material produced was called 'cloth of gold'. This was worn only by the aristocracy, particularly of southern Europe. The meeting of Henry VIII and the King of France outside Calais in 1520 was known as the field of the cloth of gold because of the many nobles wearing tunics of byssus.

The use of *Murex* snails by the Phoenicians and other people to produce tyrian or royal purple has already been mentioned. Each whelk produced only a minute amount of pigment so that an absolutely enormous number was required to dye enough material for a single garment. A contemporary industry that involves a marine invertebrate is the sponge industry. Bath sponges are still in demand despite plastic imitations, and are fished commercially in the Mediterranean, Gulf of Mexico, Carribean, Bahamas and West Pacific. Some sponge beds are planted out with sponge fragments which will grow to full size, but this is a slow procedure. The sponges are prepared by letting them dry out in the sun; the living parts shrivel and die and leave the soft horny skeleton. Sponges are one kind of marine invertebrate that may produce useful antibiotic drugs in future.

Oil and Chalk. Much of man's energy supplies are derived from oil which was formed over millions of years from the sedimented bodies of astronomical numbers of minute dead foraminiferan protozoans. The foraminiferans are *Amoeba*-like protozoans that have a coiled chambered shell punctured with holes through which fine cytoplasmic projections extend. Some foraminiferans can grow comparatively large. It is their transformed protoplasm that produced oil. Their accumulated shells have produced large chalk formations. The shells of some foraminiferans such as *Nummulites* are used by geologists as an index to rocks that may be oil-bearing. One of the best known contemporary foraminiferans is *Globigerina*. This still occurs in the oceans in large numbers and the dead sedimented bodies accumulate on the sea floor to form globigerina ooze. Unfortunately the conversion of dead animal matter into oil is a phenomenally slow process and man is at present consuming in minutes oil that has taken millions of years to mature in its chalky catacombs. This means that man will rapidly have to find alternative energy sources before he completely exhausts fossil fuels.

Another useful subterranean invertebrate, this time a living one, is the earthworm. Living in large numbers in the soil the earthworm aerates the soil and mixes it with plant-rich humus. The usefulness of these stygian gardeners in maintaining soil fertility is demonstrated by the story about the farmer who sprayed his orchard with DDT to kill insect pests. The DDT washed off into the soil where it remained for a long period, entered the bodies of earthworms as they fed and gradually accumulated in the fat body of the earthworms, eventually killing them. The ultimate result was gradual deterioration of the apple crops because the worms were no longer maintaining suitable soil conditions for growth of the trees.

Invertebrates and Disease. Man does not find all invertebrates useful, and wages constant war with many parasites and pests. Protozoans, flatworms and roundworms cause some of man's most serious diseases, and indeed often infect his domestic animals as well. Amongst the most serious of these

Cyanea lamarcki, a common jellyfish in the North Sea is blue or purple and has a painful sting.

diseases are malaria, caused by an *Amoeba*-like parasite that enters the red blood cells; sleeping sickness, caused by a flagellate protozoan living in the blood and nervous tissue; schistosomiasis or Bilharzia, caused by worms in the blood supply of the lower bowel; and hookworm disease and ascariasis, caused by roundworms in the gut. These diseases cause pain, suffering and death. They strike hardest where resistance is already lowered by famine or malnutrition and if they affect children, may maim them physically and stunt them mentally for life. Man fights a constant battle not only against the parasites themselves but against the vectors and intermediate hosts that transmit them.

Apparently innocuous Water snails transmit schistosomiasis and other fluke diseases, and present considerable control problems. Malaria and sleeping sickness are transmitted by mosquitoes and tsetse flies respectively which can become resistant to insecticides and a great deal of money is spent developing new insecticides and new control methods generally against parasites. Other parasites attack crops, the Potato root eelworm, Cotton boll worm and aphids being amongst the most notorious. Not all parasites have proved harmful to man, however. Some natural parasites of man's pests have been used to eradicate the pest with great success. This is known as biological control. Biological control was first used on a large scale to control the Cushion scale which was an accidentally introduced parasite attacking citrus fruit in Cali-

The beautiful cowrie *Cypraea histrio*. Shells of close ▷ relatives of this cowrie were used as currency in Africa until the 19th century.

fornia. The Cushion scale insect was causing an enormous amount of damage but was eventually controlled by a ladybird beetle imported from Australia in 1888. Within two years the ladybirds had virtually eliminated the parasite, and have continued to keep the parasite at a low level ever since. It is hoped that biological control may eventually be used against man's own parasites or the vectors that transmit them. Certain species of guppies are already used to control mosquitoes that transmit malaria, and there is interest in sciomyzid fly larvae that devour snails transmitting bilharzia. Another method of biological control is to sterilize a large number of males in the population so that a significant proportion of matings will be unsuccessful and lead to a population crash of the vector. A similar programme carried out on the Screw worm fly which as a larva has the abhorrent habit of burrowing into the living flesh of man and cattle, has almost eliminated this parasite from the United States.

Man's invertebrate enemies in the sea include all those venomous animals, particularly Sea wasps, that attack him or cause unpleasant reactions. Fouling organisms provide another kind of problem and are of considerable economic importance. Barnacles, mussels, Tube worms and other sedentary marine organisms cluster thickly onto the bottoms of ships and clog up the cooling intakes of power stations in estuaries. Hundreds of thousands of pounds are spent each year on research into marine paints that will discourage fouling, and into the factors that cause settling to occur. Cleaning off these organisms can also be expensive although a preliminary to cleaning the hulls of ships is to run them into a freshwater dock which causes the marine organisms to die off rapidly. Their skeletons may still have to be removed, however, and this is particularly important because the skeletons of old encrusting organisms provide a chemical stimulus encouraging new settlers to attach.

Burrowing marine organisms such as the shipworm *Teredo navalis* are also notorious pests and damage wooden pier pilings and groynes. Their effects were more widely felt in the days of wooden sailing ships, and one American zoologist has suggested that the shipworm might have been a contributory factor in the defeat of the Spanish Armada. He suggests that as the Armada's depar-

Shipworms and their burrows exposed. These are actually burrowing bivalves with the shell reduced to a cutting plate and long siphons.

ture was delayed by several months, the ships timbers may well have been eaten through by shipworm by the time they encountered the English. Even underwater steel pilings are subject to attack by burrowing Sea urchins in the United States.

A recent underwater creature that man has launched an attack against is the Crown of thorns starfish *Acanthaster planci*. Large numbers of these starfish have been seen eating coral and threatening the ecology of reefs in many parts of the tropics. One early control measure consisted of divers going down and injecting formalin into the starfish to kill them, but this was stopped when the size of the problem was appreciated, and now more effort is put into trying to understand the biology of these large spiny Sea stars than in trying directly to kill them off. Some of the information gained on them so far is most interesting. It seems that the starfish

larvae are attracted to the empty holes in coral previously occupied by living polyps so that the more devastation wrought by the starfish the greater the number of settlement sites it provides for its developing larvae. Furthermore, living coral polyps actually eat starfish larvae so this hazard too is removed by efficient grazing by the adult Crown of thorns starfish. Factors such as these may help to explain why the numbers of these starfish seem to be increasing in such a dramatic manner. Other natural predators to the adult Crown of thorns have been found. Puffer fish and Trigger fish have sharp teeth and seem to relish these not very appetizing-looking creatures. The large triton molluscs are also natural predators and in Australia the government has banned catching tritons, prized for their shells, because of the control they may exercise on the destroyers of Australian reefs.

Index

Italics are used for generic and specific names, and also to indicate pages on which illustrations appear.

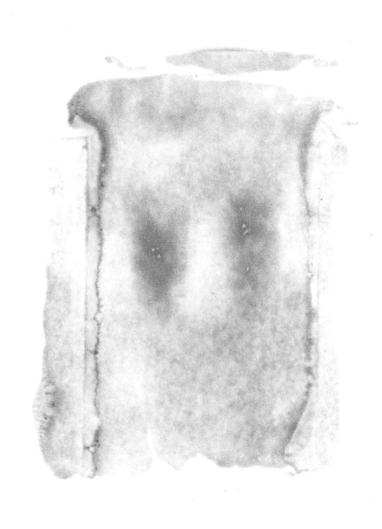